CRITICAL REASONING AND SCIENCE

Looking at Science with an Investigative Eye

M. Andrew Holowchak

University Press of America,® Inc.
Lanham · Boulder · New York · Toronto · Plymouth, UK

University Press of America,® Inc.
4501 Forbes Boulevard
Suite 200
Lanham, Maryland 20706
UPA Acquisitions Department (301) 459-3366

Estover Road
Plymouth PL6 7PY
United Kingdom

Printed in the United States of America
British Library Cataloging in Publication Information Available

Library of Congress Control Number: 2007926941
ISBN-13: 978-0-7618-3794-7 (paperback : alk. paper)
ISBN-10: 0-7618-3794-9 (paperback : alk. paper)

™ The paper used in this publication meets the minimum requirements of American National Standard for Information Sciences—Permanence of Paper for Printed Library Materials, ANSI Z39.48—1984

Dedication

To my niece, Jennifer.

Contents

List of Figures

Preface

> "The whole of science is nothing more than a refinement of everyday thinking. It is for this reason that the critical thinking of the physicist cannot possibly be restricted to the examination of the concepts of his own specific field. He cannot proceed without considering critically a much more difficult problem, the problem of analyzing the nature of everyday thinking". Albert Einstein, "Physics and Reality"

I CONSIDER MYSELF FORTUNATE to be an educator at one of the many fine academic institutions across the globe. Young men and women, many of whom have no definite plans for their life, fill the classrooms every day. After classes, they share with each other and sometimes with me their experiences of what they have learned. These experiences are about life—more specifically, about life's possibilities.

Like any good book on critical reasoning, this book too is about possibilities. If a course on critical reasoning is to do anything worthwhile, it must help and encourage students to explore possibilities through imaginative investigation of everyday-life issues. I stress "imaginative" here, as I believe that critical reasoning involves not only introduction to the proper norms of thought, but it also invites students to develop and expand their imagination.

Why however have I written a book on critical reasoning and *science*? The answer is simple. Critical reasoning and science is an especially underdeveloped area. While critical reasoning at many academic institutions is one of the most important desiderata of general education, critical investigation of science, at both theoretical and practical levels, is generally ignored. I find that to be both astonishing and perplexing. The perception is perhaps that, with so much science practiced at institutions, students will learn about science by taking a course or two in particular sciences. While most students do take some courses in particular sciences, they go away without a clear grasp of what science is, how it is practiced, and how to evaluate it.

Scientific investigation influences our world profoundly each day with new discoveries, theories, and hunches about the world and our place in it. Astrophysicists study deep-space objects to learn about the history of the universe. Epidemiologists seek cures for deadly diseases. Psychologists search to find the most fundamental motives of human behavior. Sociologists look for tendencies in behaviors of animal groups. Pharmacologists market new drugs to ease and prevent human suffering. Evolutionary biologists trace out the mechanisms for the development of our species—*homo sapiens*. While such things go on every day in science, most of us feel a certain estrangement from scientific inquiry.

This book is an attempt to eliminate or at least diminish the feeling of estrangement that many feel toward science. It is divided into two main parts—a critical look at philosophical issues related to science and a critical look at the practice of science.

Since it is assumed that the material covered throughout is new or foreign to most readers, I introduce it in short, digestible units—called "modules". The idea behind these modules is to give students no more information in each module than they can assimilate at a certain time. Every module includes at least one "text box" to give students background information on the history of science or provide additional information on topics addressed in the book that may be of interest to students. In general, these text boxes give the book a historical storyline, from the early Greeks to present-day science, which ties text material to the practice of science. At the end of each module, I include key terms, text questions, and text-box questions. Some of the modules, especially those in the second part of the book on philosophy of science, are wholly devoted to practice exercises. Three of the modules in the third part of the book, devoted to the practice of science, contain nothing but practice exercises. Overall, there are 26 modules.

Part I, "Introduction: Science and Critical Reasoning", begins with four introductory modules that discuss the relationship of science and critical reasoning. Module 1 concerns the nature of science. Module 2 looks at critical reasoning. Module 3 distinguishes between inductive and deductive reasoning and then lists two commonly used inductive arguments and two commonly used deductive arguments in science. Finally, the fourth module ties together the material in the first three modules by an analysis of both normative thinking in science and scientific discovery through imaginative thought.

Part II is entitled "Philosophy of Science with an Investigative Eye", which comprises three sections—one, which analyzes the relationship between science and truth; another, which takes a philosophical look at causation; and the last, which analyzes the relationship between science and progress. In these eight modules, my aim is to introduce some philosophically technical material in a relatively straightforward and simple manner. The issues are theories of truth, realism versus antirealism, causal methods and causal fallacies, the possibility of knowledge, and logical and historical accounts of science.

Part III is entitled "Scientific Practice with an Investigative Eye", which comprises three sections—one on theoretical models, one on statistical hypotheses, and one on causal hypotheses—and 12 modules. The main purpose of these sections is to acquaint students with episodes of science to which they are exposed every day through newspapers like the *New York Times*, magazines such as *Time* and *Science News*, and internet sources such as *World Science*. In these sections, I introduce analytic tools that enable students to read and evaluate episodes of science critically. Three of the modules specifically illustrate just how to use such tools; three other modules are entirely devoted to practice exercises. Throughout these modules, key terms like "model", "theory", "hypothesis", "standard deviation", "cause", and "correlation" are fully examined in a manner that is, I hope, accessible to all introductory-level students.

The last part, "Science and Its Pretenders", comprises two modules that discuss criteria that may be used to distinguish science from non-science. Of these two, the final module gives practice exercises.

Overall, *Critical Reasoning and Science* is unique in aim and functionality, as it is the first book to offer students a critical approach both to the philosophy and to the practice of science, and it does so in a user-friendly manner. There are, however, many fine texts and anthologies on the philosophy of science. John Losee's *A Historical Introduction to the Philosophy of Science* and Anthony O'Hear's *An Introduction to the Philosophy of Science* are two examples. On the other hand, there are too few books aimed at helping students gain critical insight into the practice of science. Of these, Ronald Giere's *Understanding Scientific Reasoning* is, I think, the best. Many of the evaluative methods I employ in Part III are derived, at least in part, from Giere's groundbreaking work. Thus, the aim of *Critical Reasoning & Science* is to give students tools for analysis of science by looking critically at both its practice and its philosophy.

Before ending, I must acknowledge certain debts. First, there is my debt to the exceptional instructors and friends that I was privileged to have as a student while in the Department of History and Philosophy of Science at the University of Pittsburgh. Second and most importantly, there is my debt to the very many great and courageous men and women over the centuries, who have contributed to the development and advance of science and philosophy through imagination and single-minded persistence in the face of innumerable obstacles. Last, of course, there is the debt I owe my students at Kutztown University, who patiently sat through early drafts of this material and offered numerous beneficial criticisms along the way.

PART 1

INTRODUCTION

SECTION ONE

Science & Critical Reasoning

Module 1
What Is Science?

"The size of a man's mind ... is to be measured, in so far as it can be measured, by the size and complexity of the universe that he grasps in thought and imagination". Bertrand Russell, "Adventures of the Mind"

As VITAL SCIENCE IS TODAY to human flourishing, it is also one of the most misunderstood human practices. On the one extreme, many people trust scientists' utterances as if they were divine pronouncements, incapable of being false. On the other extreme, many people distrust just about everything scientists say. Both positions are, of course, absurd. Each is the result of human ignorance. What is common to each is an almost complete misunderstanding of what it is that scientists really do.

Not knowing what science is certainly makes it impossible to evaluate rationally and responsibly the practice of science. Thus, this book is principally an attempt to make science critically accessible to introductory-level college students.

Mythology and Science

In ancient times, people passed their time through "story-telling" (Greek, *mythoi*). These myths were not only entertaining; they also offered people comfort and a sense of control over their lives by humanizing the world around them. Sigmund Freud writes of this first attempt to gain some element of control over nature:

> A great deal is already gained with the first step: the humanization of nature. Impersonal forces and destinies cannot be approached; they remain eternally remote. But if the elements have passions our own souls, if death itself is not something spontaneous, but the violent act of an evil Will, if everywhere in nature there are Beings around us of a kind that we know in our own society, then we can breathe freely, can feel at home in the uncanny and can deal by psychical means with our senseless anxiety. We are still defenceless, perhaps, but we are no longer helplessly paralysed; we can at least react. Perhaps, indeed, we are not even defenceless. We can apply the same methods against these violent supermen outside that we employ in our own society; we can try to adjure them, to appease them, to bribe them, and, by so influencing them, we may rob them of a part of their power. A replacement like this of natural science by psychology not only provides immediate relief, but also points the way to a further mastering of the situation.[1]

Through myths, people justified their ritualistic practices, paid honor to gods, systematized and even tamed the world around them.

Whereas myths perhaps have their roots in our "senseless anxiety", science has its roots in our sensible curiosity about how nature works. As Aristotle said some 2400 years ago, "In all natural things, there is something wonderful".[2] The delight, Aristotle adds, comes principally not through the use to which we can put such knowledge, but rather in the mere having of such knowledge. However, unlike mythology, scientific understanding comes not through humanizing nature, but through dehumanizing it.

Perhaps the earliest practice of science comes with the Babylonians and Egyptians. Both developed sophisticated celestial charts and arithmetic schemes for navigation, for establishing a fixed lunar calendar, for architectural placement of buildings, and for predicting the future. The Egyptians also developed a practical science of geometry for purposes of irrigation in farming. Neither the Babylonians nor the Egyptians, however, developed a science that aimed at pure understanding, independent of its practical benefits.

True science, then, began with the Greeks. These early thinkers did not seek to explain observable phenomena by recourse to the irrational or supernatural. Instead, in time, they appealed to their own rational capacities and the regularities in nature itself. With the Greeks, supernatural explanation began to give way to natural explanation.

Science, then, comes from a curiosity to know the *why* of natural things—a curiosity that is, quite possibly, uniquely human. *Science is*, we may say, *a systemized attempt to learn about ourselves and the world around us that uses experiment and observation.*

Greek Monism

The first attempts at systematic and rational understanding of the cosmos and the motion and change within it were *monistic* (implying that there is some primary reality (1) that underlies all change or (2) that is something out of which all things come) and materialistic. The early Milesians in the sixth century B.C.—Thales, Anaximander, and Anaximenes—sought to reduce the flux of all visible phenomena to one sort of underlying, unchanging stuff. For Thales, this stuff was water. He also posited that all things were full of gods. Anaximander, who is believed to have been a younger contemporary of Thales, said that the underlying stuff was a boundless, indefinite sort of matter (Gr., *apeiron*) out of which all things came and to which all things necessarily returned. The third Milesian, Anaximenes, perhaps as a compromise between the boundless *apeiron* of Anaximander and the crude corporeality of Thales, stated that air was the primal stuff. Through rarefaction and condensation, air comprised all visible things. While fire was thought to be rarefied air, water, earth, and stone were air at increasingly greater levels of condensation.

Heraclitus of Ephasus (d. c. 480 B.C.) argued that the flux of observable phenomena was itself real and perhaps the only thing that was real. This flux he tended to identify with fire—a material element that was itself ever changing, yet ever the same. A wise person, for Heraclitus, was one whose soul was dry.

In stark contrast to Heraclitus and the early Milesians, Parmenides of Elea (c. 515 B.C. to c. 450 B.C.) maintained that rational reflection about the true nature of what is real imposed certain self-evident standards on this reality. In his poem *The Way of Being*, Parmenides asserted that "What-is is" and "What–is-not is not". What-is, his ultimate reality, was said to be (1) one, (2) ungenerated, (3) indestructible, (4) eternal, and (5) unchanging. Thus, Parmenides was the first true rationalist.

The Aims of Science

Science is a discipline that employs methods that enable us to gain some understanding of how the world really is independently of us. As such, it gives us information that is not only intellectually satisfying but also useful for improving the quality of our lives and for solving social and political problems. On the one hand, knowing that different types of stars have different life-cycles and different life-expectancies is of little practical assistance in our lives, but such things do contribute to our overall understanding of the universe and this in itself, to many, is well worth the effort of study. On the other hand, so many of the technological advances of scientists—from mechanical or computerized kitchen items to pharmacological products at the local drug store—are an undeniable part of our everyday lives, for better or worse.

What are the aims of science? Let me illustrate by contrasting science with mythology.

First, science allows us to have some measure of *command* of our lives and the world around us. In science, one of the best examples of management comes through health sciences like medicine and nutrition science. Discoveries in health sciences allow people to live healthier and longer lives. Research on the antioxidant capacities of certain foods and vitamins allow people to retard and check free-radical degeneration of living tissue in bodies. Such research marks important advances in our understanding of aging and cancer. Pain killers allow many with mild forms of arthritis to live relatively pain-free lives. Improvements in surgical techniques over the years give people improved bodily movement. The slow development of psychology has given us greater understanding of the causes of deviant and pathological behavior.

In contrast, mythological accounts of phenomena merely give us the illusion of command. Myths allow us to see ourselves in control of events, only because, through myths, we construct an account of these events in human terms. This sort of constructive control is entirely subjective. For example, ascribing to the god Poseidon rule over the seas gave early Greeks a perception that they could propitiate him through prayers and offerings, designed to confer his blessings before voyages or during sea battles, and calm or keep calm the seas. If the seas were calm for a voyage, Poseidon was duly propitiated. If not, then either their prayers and offerings were insufficient or the god was angry. In short, for believers in constructive control, the way things really are in reality is irrelevant to one's perception of reality. One's beliefs need not conform to reality; reality needs to conform to one's beliefs.

Science also enables us to *explain* phenomena. Consider two explanations to the question, "How did human beings come about?"

One Babylonian myth gives us the following account. There were three primordial water deities: Apsu (Sweet-Water Ocean), his wife Tiamat (Salt-Water Ocean), and their child Mummu (Mist). In time, these primordial deities gave rise to many other deities. These younger gods, however, were especially rambunctious and noisy. Not being able to handle this noise, Apsu and Mummu plotted to kill them. One of these rambunctious offspring, Nudimmud, found out about the plot and slew Apsu and imprisoned Mummu. Taimat was subsequently convinced by her consort Kingu to seek vengeance through an out-and-out generational war among the older and younger gods. Nunimmud, then, sought counsel among the offspring and the youngest and boldest of them, Marduk, proposed to lead the fight, if he was made head of the gods. The counsel of younger gods agreed and there ensued a colossal war among the gods. In the battle, Marduk split Tiamat's skull and her blood was then carried south by the North Wind. Marduk then created the universe, both sky and earth, by dividing her body. He next created the stars, an abode for the gods, the moon, and the sun. Last, he killed Kingu and, from him, created man to serve the gods.

Of course, nothing can rival a drama of a fight between the gods. Nonetheless, the scientific answer to the question is, I believe, it is every bit as fascinating. It goes roughly like this. Over the course of many millions of years, human beings have evolved into bipedal, rational creatures from simple life-forms that presumably developed out of complex molecules. We have not yet discovered how simple life-forms come about from complex molecules, but we do know that all living organisms are in the process of evolutionary changes. How do we know this to be true?

Here, unlike mythological accounts, scientific accounts appeal to evidence that is accessible to other scientists and non-scientists alike. That evolution of some sort is true we know from direct observation and records of fossils. First, in people with the AIDS virus, physicians literally watch the virus mutate within patients as a response to various medicines given to them. Second, genetic similarities between humans and other primates, such as chimpanzees and apes, are too remarkable to be a product of chance and support evolution from a common ancestor. Last, stark similarities of fossilized remains of early primates with existing plants and animals today strongly suggest an evolutionary link.

General Theory of Relativity Confirmed

Einstein's General Theory of Relativity, put forth in 1915, was a new account of gravity that was inconsistent with Newton's theory. According to both theories, light should bend as it passes around a massive object. For Newton, this is because *light is made of particles and all bodies with mass are attracted to each other*. For Einstein, *light is without mass, but all massive bodies warp the space around them, thus causing the light that passes by such an object to bend.* Relativity theory predicted that light, passing by our sun, would bend about 1.75 arc seconds—much more than predicted by Newton's theory of gravity. As a test of sorts, Einstein himself envisaged a scene where a star, slightly obscured by the edge of the sun, S, would be seen by someone at position O to be just outside of the edge of the sun, due to the warping of space around the sun.

This, of course, was a scenario that was impossible to observe due to the brilliance of the sun. However, such an observation could be performed during an eclipse of the sun. Thus, after World War I, astronomer Arthur Eddington traveled to the island of Principe, off the coast of West Africa, to observe the eclipse. Most of the 410 seconds of total eclipse was marred by cloudiness, but Eddington did manage to get one crucial photograph of a key star just outside of the sun's edge within the last 10 seconds. When he compared the position of this star during the eclipse to its position in the sky when the sun was not around, Eddington measured the perturbation to be about 1.61 arc seconds (exaggerated below). This was very much closer to Einstein's prediction than that of Newton and was hailed as confirmatory evidence of Einstein's theory of gravity, though it is far from clear today that the measurements were as precise as claimed to be.

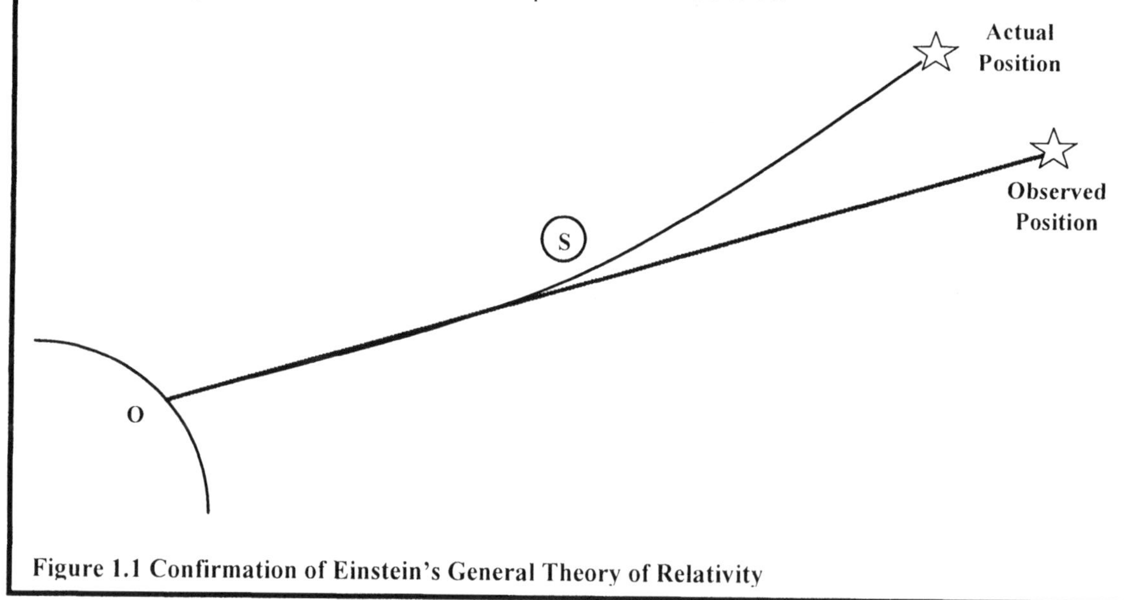

Figure 1.1 Confirmation of Einstein's General Theory of Relativity

Third, science enables us to *predict* phenomena—something that was not a part of mythology. Prediction is a tremendously significant part of scientific practice, for without it, scientific hypotheses would be no more interesting that hunches or guesses. Prediction allows us to put scientific hypotheses to the test and testing is, in large part, what separates scientific claims from claims of other sorts. For instance, according to the generalization of relativity theory, $E = mc^2$ (where "E" is energy, "m" is mass, and "c" is the speed of light), light, though without mass, will itself bend as it passes by massive objects, as massive objects have a pronounced warping effect on the space around them. It was commonly held at the time Einstein put forth his General Theory of Relativity in 1915 (see Figure 1.1) that space itself was empty and incapable of being affected in any way by matter. A predictive consequence of relativity theory was that any star, situated directly behind the sun, would appear displaced during an eclipse, since its light on its way to the earth will bend as it passed the sun. May 29 of 1919 offered just such a test. Arthur Eddington oversaw two expeditions—one, in Brazil; the other, off the coast of West Africa. Both expeditions confirmed Einstein's predictions and earned the scientist instant celebrity status.

Finally, though this issue is hotly contested by philosophers of science (as will become evident in upcoming modules), science aims at knowledge of the way things are or truth, while the aims of mythology are many—explanation, entertainment, ritualistic initiation, and even rationalization.

Greek Pluralism

Other Greek philosophers realized that the constant flux of things could more readily be explained less economically but more viably by a metaphysics that rejected monism. Pluralism was born.

Empedocles of Acragas (c. 493-c. 433 B.C.) asserted that the cosmos was a plenum and that there were four fundamental "roots" or underlying principles—fire, air, water, and earth—and two motive (perhaps material) forces—one that brought things together (Love) and one that tore things apart (Strife)—all of which worked together in the eternal cycle of being.

Pythagoras of Samos (b.c. 570 B.C.) and his followers took scientific and philosophical practices and blended with them religious ritual, magic, and mysticism. Pythagoreans obeyed precepts like "Do not eat beans" and "Do not stir fires with iron". They also believed in the immortality of the soul. Much of this philosophical cult was a prescription for a way of life to purify the soul through a mathematical understanding of nature. Pythagoreans practiced science not for practical reasons, nor to quiet an innate curiosity. Science was a way of life in preparation for a proper life after death. (I return to the Pythagoreanism later.)

Anaxagoras of Clazomenae (c. 500-c. 428 B.C.) posited "there is a portion of everything in everything". Even the most uniformly fleshy piece of flesh of the smallest conceivable size had water, hair, tooth, bark, and all other things in it—though flesh predominated. Moreover, he maintained that there was no smallest part of matter. However small some bit of matter might be cut, there would always be another cut that could be made and this even smaller piece would have all things in it too. At the cosmic level, Mind was said to move all things, though in mechanical fashion.

Democritus of Abdera (c. 460-c. 370 B.C.) tackled the problem of change by asserting the existence of small, invisible, and unalterable uncuttables (Gr., *atoma*). For Democritus, there are an infinite number of atoms of differing sizes and shapes that whirl about in the vortex of empty space. As they swirl, they come together to fashion visible objects and their parts.

Critical Reasoning & Science

To talk of the origins and aims of science is not the same as to say just what science is. This is where the critical-reasoning part of *Critical Reasoning and Science* comes into play. The modules that follow are attempts at developing and honing a critical attitude toward science and its practice, both from within and without.

From without, adopting a critical attitude toward science is adopting a philosophical attitude.

To have a philosophical attitude means to step back from the practice of science and see just what scientists are really doing. That, in essence, is doing philosophy of science, which asks *meta-scientific questions* such as the following:

* What precisely is science?
* Can science enable us to know the world around us?
* Does science aim at truth or is it merely a problem-solving practice?
* Is there a method that defines scientific practice?
* What is the relationship between science and society?
* In what sense, if any, is science progressive?

From within, adopting a critical attitude toward science is at once to immerse oneself in the practice of science. Scientists certainly have all sorts of motives for their research. A scientist may work on cancer because her husband has cancer. Another may study evolutionary biology due to frustration with his religious upbringing.

In spite of the underlying motives of scientists, one of the express aims of science is *value-neutrality*. Value-neutrality means that the motives behind scientific research ought not to show themselves at all in the research or the end product. For instance, an evolutionary biologist, who boasts that his latest research will be the death knell of creation science, is probably not doing objective research and it is unlikely he will get his research published.

Science, then, is a matter of putting forth hypotheses about the world, testing these hypotheses, listing all relevant data concerning the tests, and then drawing up, in value-neutral terms, all the implications of such research.

Looking ahead, some of the critical-thinking goals in subsequent modules will be practical and philosophical. They are as follows:

ACCESSIBILITY: *To enable us to gain a fuller grasp of what scientists do.*
UNDERSTANDING: *To measure the extent to which science gives us knowledge of natural phenomena.*
CRITICISM: *To develop a critical apparatus for evaluating and judging scientific practice.*
DEMARCATION: *To establish criteria that mark off genuine science from pseudo-science.*
DIFFERENTIATION: *To show how science differs from other disciplines and social practices like religion, philosophy, and mythology.*
INDISPENSABILITY: *To show the significance of scientific reasoning in our everyday lives.*

KEY TERMS

science
meta-science
scientific command
truth
pluralism
value-neutrality
scientific prediction
scientific explanation
monism

TEXT QUESTIONS

* Using examples of your own, give illustrations of each of the main aims of science. Why is value-neutrality important?
* Explain how scientific explanation differs, at least in intent, from mythological explanation or other forms of explanation that do not appeal to experience?

TEXT-BOX QUESTIONS

* How did early Greek pluralism mark an advance over early Greek monism?
* How did Eddington's 1919 observations prove to be a crucial test that decided between Einstein's General Theory of Relativity and Newton's theory of gravity?

---

1 Sigmund Freud, *The Future of an Illusion,* trans. James Strachey (New York: W. W. Norton & Company, 1961), 20-21.
2 *Parts of Animals* I.5 (645a16). My translation.

Module 2
What Is Critical Reasoning?

"All our dignity consists, then in thought. By it we must elevate ourselves, and not by space and time which we cannot fill. Let us endeavour then to think well; this is the principle of morality". Blaise Pascal, *Pensees* #347

Aims of Critical Reasoning

Geneticists study the hereditary transmission of traits and diseases to gain a fuller understanding of human nature. Political statisticians, seeking causally relevant information, examine trends in voting populations to give themselves clues for campaigning strategies. Observational astronomers, guided by computer-aided technology, look back in time through telescopes to search for the origin of the cosmos. Clinical psychologists use free association and dream interpretation to find early trauma that is causally linked to pathological behavior. Archeologists analyze remnants of tools at an ancient excavation site to piece together the manner of life of a long-lost civilization. Today, more than any other time in the history of the world, science is recognizably a great part of the lives of people.

Yet what are the aims, principles, and methods of science? To what extent can science justify its claims? How do observation and experiment come into play? These questions and numerous others give us pause to ask, "What precisely is science?" Yet before we take up these questions, a few words on critical reasoning are needed.

There are two components to critical reasoning. First, to engage in critical reasoning, one must have a *critical attitude* toward all issues that are presently unsettled by reasoned discussion. A critical attitude on such issues entails open-mindedness, an admission of fallibility, and a commitment toward intellectual integrity. *Open-mindedness* and *fallibility*, of course, go together. When rationally contested issues are unsettled, it is absurd to be close-minded and to begin from a position of infallibility. This is a sure way to prevent debate from moving forward in an effort to decide the issue. *Intellectual integrity* is a commitment to seek out evidence and to follow it wherever it goes—that is, to allow the evidence to suggest the right conclusion, instead of working under preconceived biases or prejudices.

Comprising open-mindedness, an admission of fallibility, and a commitment to intellectual integrity, a critical attitude aims at personal stability, conflict resolution, and success in evaluating and attaining goals. One with a critical attitude is troubled by inconsistencies in thought and strives as fully as possible to resolve internal inconsistencies. The result is *personal stability*. One with a critical attitude is also troubled by interpersonal conflict. Though differences between humans are inevitable, these need not be taken as signs of inescapable irresolution. *Resolution of conflict* is always possible. With a critical attitude, differences of perspective on a volatile issue become an opportunity for getting at the truth or, at least, for having a clearer perspective of the issue. Last, one with a critical attitude will probably have greater *success in evaluating and achieving goals*. One who customarily deliberates before doing most things is one who is in a good position to know what goals are worth having and one who has a greater likelihood of attaining those goals.

Socrates the Inquirer

One of the most controversial figures in human history was the Athenian philosopher Socrates (469-399 B.C.)—the centerpiece of so many of Plato's dialogues. Plato depicted Socrates as a truly remarkable person, who had exemplary reasoning skills, a mordant wit, a penetrating intellect, and a singular devotion to the pursuit of truth. Socrates, strolling around the Athenian marketplace, would spend his days in philosophical conversation in a quest for knowledge, which he equated with virtue. Though he claimed to have no knowledge, he would converse with anyone willing to discuss knowledge—that is, the particular virtues like courage, self-control, justice, piety, and wisdom. His manner of inquiry, according to Plato, was dialectical: He would ask questions of fellow inquirers, usually young men, so as to elicit a definition of or an opinion on a particular virtue. Socrates would then attempt to refute the definition. His interlocutors would then refine the definition or come up with a new one. This method of collaborative inquiry would continue until such time that the interlocutors' ignorance was fully disclosed. With their ignorance disclosed, the interlocutors would leave off inquiry frustrated. Some would pledge to pursue the issue continually until it was resolved. Others, convinced that they knew what they did not know, would rush off angrily.

We cannot really know that Plato's depiction of Socrates is accurate. Two other contemporary writers offer different accounts. Xenophon's Socrates never professes to be ignorant. On several occasions in the *Memoirs*, for example, he has Socrates sanction a moral view that Plato's Socrates considers and rejects in *Republic*—that it is right to help one's friends and harm one's enemies. The playwright Aristophanes also refers to Socrates in four of his surviving works. *Clouds* is entirely about Socrates and shows him to be a charlatan and buffoon. Aristotle, writing years after the death of Socrates, gives additional insight into the character of Socrates in comments scattered throughout his vast corpus. Overall, though a perfectly consistent picture of Socrates is impossible, what we can say with certainty is that he was a truly remarkable person, with a singular devotion to the pursuit of wisdom.

The second component to critical reasoning involves *adoption of normative criteria of thought*. Normative criteria are rules of proper thinking—not rules that describe how we do in fact think, but rules for how we ought to think. No one, for instance, who knows that he has a one-in-1000 chance of winning a certain lottery, would be rational to play that game with the complete expectation of winning. Anyone arguing that the complete expectation of winning is justified would certainly not be rational. Thus, critical reasoning involves the adoption of criteria for rational thinking and these criteria are generally assumed to be discipline-independent—that is, they apply to thinking in any discipline whatsoever.

Before we can apply the evaluative criteria in critical thinking to arguments, we must first discuss arguments.

Arguments and Their Parts

Arguments are commonly understood as verbal disputes between parties—often ones that are heated—where each is trying to convince the other of the truth of some point. In such exchanges, the methods of persuasion are virtually limitless. Each party may choose reasons, force, pity, or even trickery to get a point through to another.

For the purposes of critical reasoning, this is both too narrow, in one sense, and too broad, in another. It is too narrow in that persuasion is merely one of the aims of critical reasoning. There are others. It is too broad in that, when persuasion is the aim, the means of persuasion must not be rhetorical, but based exclusively on evidence or reasons.

Preliminary Definitions

As philosophy is a discipline that essentially aims at clarity, it is fitting to begin with some preliminary definitions.

An ARGUMENT is a collection of statements whereby evidence in the form of at least one statement (the premise) is given in support of another statement (the conclusion).

A PREMISE is a statement that is given as evidence for another statement in an argument.

The CONCLUSION is a statement that purported follows from the evidence in an argument.

Example:

> [1]<u>No theory that fails to yield a testable prediction can be called scientific</u>. [2]The reason for this is that testability is a necessary component of a scientific theory.

Here statement one (the conclusion) purportedly follows from statement two (the premise), which is given as evidence for statement one. The relationship between premise(s) and the conclusion can be depicted as follows:

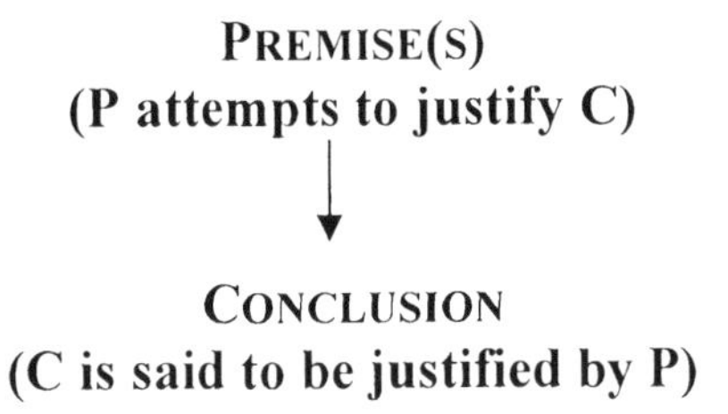

In summary, premises are put forth in order to justify a conclusion—to show that it is true—while the conclusion in the right sort of argument is justified by the premises. Overall, there must be at least two statements for an argument to exist. No claim given by itself, however contentious it may be, can be an argument.

Arguments vary greatly in their complexity. They can be simple, where one premise is given to support a conclusion, or they can be complex, sometimes horribly so—containing several smaller arguments within them, in chain- or branch-like fashion, to establish a conclusion. For instance, Charles Darwin's *Origin of Species* can be seen as one frightfully lengthy argument to show that organisms are not fixed, but change very slowly over time, chiefly through the mechanism of Natural Selection. This issue of complexity need not concern us here.

Two Reason that We Argue

We give arguments in an attempt to PERSUADE OTHERS OF THE TRUTH OF SOME STATEMENT through an appeal to evidence and reasons in support of that statement, not appeals to pity, slander, force, or any other irrational means of persuasion.

Chief World Systems

Premises support conclusion well: _____ Premises do not support conclusion well: _____

3. Americans are not incapable of understanding science. On the contrary, no group of people that has accomplished what the people of the United States have accomplished can be anything but highly competent in science. Therefore, if science illiteracy shows up, the explanation must lie elsewhere. Paul Sukys, *Lifting the Scientific Veil*

Premises support conclusion well: _____ Premises do not support conclusion well: _____

4. Of all studies, this one (i.e., astronomy) especially prepares men to be perceptive of nobility both of action and of character. When the sameness, good order, proportion and freedom from arrogance of divine things are being contemplated, this study makes those who follow it lovers of this divine beauty and instils, and as it were makes natural, the same condition in their soul. Ptolemy, *Almagest* I.1

Premises support conclusion well: _____ Premises do not support conclusion well: _____

5. Time is warped so drastically on a neutron star because the star, while possessing the mass of the sun or more, is nevertheless compressed to a radius of just a few kilometres. The stronger the gravity at the surface of an object, the more time is slowed, or stretched. Paul Davies, *About Time*

Premises support conclusion well: _____ Premises do not support conclusion well: _____

6. In the course of the long time down to the present, though, with a multitude of observers having observed a vast array of things, many things have been found to have been observed in vain. For this reason, it is now history, which is most useful, while earlier it was one's own perception. Galen, *An Outline of Empiricism,* VI

Premises support conclusion well: _____ Premises do not support conclusion well: _____

7. Anarchism helps to achieve progress [in science] in any one of the sense one cares to choose.... It is clear, then, that the idea of a fixed method, or of a fixed theory of rationality, rests on too naïve a view of man and his social surroundings. Paul Feyerabend, *Against Method*

Premises support conclusion well: _____ Premises do not support conclusion well: _____

8. He who has eyes to see and ears to hear becomes convinced that mortals can keep no secret. If their lips are silent, they gossip with their fingertips; betrayal forces its way through every pore. Sigmund Freud, *Fragment of an Analysis of Case Hysteria*, 1905

Premises support conclusion well: _____ Premises do not support conclusion well: _____

9. They [plants and animals] are all impelled by a powerful instinct to the increase of their species, and this instinct is interrupted by no reasoning or doubts about providing for their offspring. Wherever therefore there is liberty, the power of increase is exerted, and the super-abundant effects are repressed afterwards by want of room and nourishment, which is common to animals and plants, and among animals, by becoming the prey of others. Thomas Robert Malthus, "An Essay on the Principle of Population"

Premises support conclusion well: _____ Premises do not support conclusion well: _____

10. No matter how large the environment one considers, life cannot have had a random beginning. Troops of monkey thundering away at random on typewriters could not produce the works of Shakespere, for the practical reason that the whole observable universe is not large enough to contain the necessary monkey hordes, the necessary typewriters, and certainly the waste paper basket required for the deposition of wrong attempts. The same is true of living material. Fred Hoyle, *Evolution from Space*

Premises support conclusion well: _____ Premises do not support conclusion well: _____

Module 3
Inductive & Deductive Reasoning

"'There is no use in trying' said Alice; 'one can't believe impossible things'.
'I dare say you haven't much practice', said the Queen. 'When I was your age, I always did it for half an hour a day. Why, sometimes I believed six impossible things before breakfast'".
Alice's Adventures in Wonderland, Lewis Carroll.

ON MOST PHILOSOPHICAL VIEWS of how science works, both inductive and deductive reasoning have an important role. Yet there are extremes. On the one hand, extreme inductivists see science as essentially and principally driven by inductive arguments. On the other hand, extreme deductivists, like falsificationists, are unwilling to grant induction any significant role in science and argue that science is essentially a deductive practice.

There is more to say about both inductivism and deductivism in upcoming modules. For now, let us get familiar with inductive and deductive arguments.

Inductive Arguments

Let us begin with a definition of "inductive argument".

An INDUCTIVE ARGUMENT is an argument in which the premises attempt to support its conclusion strongly, though not absolutely.

An inductive argument is *strong*, when it fulfils the Condition of Relevant Support in Module 2.

An INDUCTIVELY STRONG ARGUMENT is an inductive argument whose premises, when assumed true, give one more reason than not to believe that the conclusion is also true.

The right sort of inductive argument, one that is *cogent*, will be one that fulfils both Criteria of Acceptance in Module 2: It will have all of its premises true and these premises will give, roughly speaking, good reason to believe that its conclusion is true.

An INDUCTIVELY COGENT ARGUMENT is an inductive argument whose premises, when actually true, give one more reason than not to believe that the conclusion is also true.

In general, the more relevant evidence there is in the premises, the more reason there is for accepting the conclusion.

Scientists use many types of inductive arguments in many ways. We look at two of the most common types used in science below: inductive generalization and argument from analogy.

Aristotle on Knowledge

At *Metaphysics* A.1, Aristotle traces the path of the acquisition of knowledge from perception to knowledge and technical science. Knowledge begins with perceptions, perceptions lead to memory, and many memories of similar perceptions form experience. From experience, we arrive at two kinds of scientific understanding: knowledge and technical science.

> And experience seems nearly the same as knowledge and technical science, and knowledge and technical science come about from experience. For experience fashioned technical science, as Polus rightly says, while lack of experience fashioned chance. And technical science occurs whenever, from many ideas of experience, a single universal notion occurs from similar things. For having a notion that this thing cured Callias (who was sick with a certain illness) and Socrates and similarly many others, each in its own way, is a matter of experience. But having a notion that this thing cures all such men, designated as one class, who suffer from a certain illness—for example, the phlegmatic or choleric or those with burning fever—is a matter of technical science.

First, though they involve different objects, technical science and knowledge both involve a grasp of what is universal; experience does not. Next, technical science and knowledge are built ultimately from an elaborate process of collecting and storing perceptions. The principles of both, then, are based on some form of gathering particulars that is similar to inductive generalization.

Yet, Aristotle cautions, people with experience are more successful than people with rational understanding who have no experience. The reason he gives is "experience is an understanding of particulars, while technical science is an understanding of what is universal, and all deeds and productions are concerned with particulars". Physicians, he adds, cure particular men, like Callias and Socrates, directly; man, in general, they cure only accidentally. Therefore, one who knows only what is universal and gives a rational account without experience will often err in treatment. For therapy, he says, concerns what is particular.

Nevertheless only technical men possess wisdom. Aristotle elaborates:

> Still, we think that knowing and expertise belong to technical science rather than experience, and we suppose that technicians are wiser than men of experience, so that, for all men, wisdom attends upon knowing instead. And this is because some men (technicians) know the cause, while others (non-technicians) do not. For men of experience know the fact, but they do not know the reason why; while others discover the reason why and the cause.

The sign of knowledge, he asserts, is the ability to teach. Because of this, in a sense, "technical science is knowledge, rather than experience". That is why theoretical science is the most esteemed.

Thus, for Aristotle, there are three levels of scientific investigation. First, there is *knowledge* in the true sense of the word. Starting from true premises that involve many experiences of things invariable, we gradually develop universal principles that are certain. From these, we deduce true conclusions. Dealing with what is universal, knowledge is teachable. Knowledge, then, is pure theoretical science. Second, there is technical science. *Technical science*, like knowledge, begins with premises ultimately derived from many experiences of things. Yet unlike knowledge, these things, being items of manufacture, are variable. Thus, its premises are, at best, likely to be true and its conclusions are suspect and void of certainty. Nevertheless technical science, as productive science, involves causes (i.e., suspected causes, since its objects are variable) and is thus closer to real knowledge than experience. It, too, involving what is universal, is teachable. Last, *experience* is exclusively about particulars. Not involving what is universal, it is a knack for doing the right thing that is guided by a keen awareness of particular circumstances and past observations. Experience may be a useful guide to successful scientific practice, but it does not involve causes or what is universal and, thus, is not teachable. Most importantly, with no regard for what is universal, consistency is not an issue: It is quite possible that what has once worked under a certain set of circumstances will not again work. This is scientific knack.

Inductive Generalization

One of the most commonly used types of inductive argument is inductive generalization.

An INDUCTIVE GENERALIZATION is a type of inductive argument that goes from particular premises to a generalized conclusion.

In an inductive generalization, a certain amount of information about some subject is gathered and this information itself suggests a conclusion with universal scope. For instance, consider this argument about lycopene below, where number "n" of rats with a prostate condition, are given lycopene and each shows improved prostate health.

Example one:

> Rat_1, given lycopene, shows improved prostate health.
> Rat_2, given lycopene, shows improved prostate health.
> ...
> Rat_n, given lycopene, shows improved prostate health.
> So, all rats, given lycopene, will show improved prostate health.

We can simplify the argument by condensing the evidence in the premises as follows:

> Number "n" of *observed* rats, given lycopene, show improved prostate health.
> So, *all* rats, given lycopene, will show improved prostate health.

Example two:

> 67% of the times I've gone to Keratokambos, I've gotten ill.
> So, 67% of *all* the times I'll ever go to Keratokambos, I'll get ill.

In example two, the generalization in the conclusion is a *statistical generalization* and the premise is statistical in nature. As later modules will show, this type of inductive generalization is extraordinarily important in both statistical and causal models. In example one, the conclusion is a *universal generalization*, gathered from a finite amount of particular data in support of it.

General form:

> X% of *observed* Ss has attribute α.
> So, X% of *all* Ss has attribute α (X being from 0% to 100%).

(Here "X" can be any number between zero and 100. Note that the conclusion becomes universal, when it is at one of these two extremes; otherwise, it is statistical).

When it is reasonable to accept the conclusion of an inductive generalization as true? To answer this question, we need to look at two important standards for assessing the strength of an inductive generalization: the principle of sufficient variation and the principle of sufficient size.

does not have, and these could be relevant and good reasons for T *not* having δ.

Analogical arguments are often expressed more economically. Consider the following:

Rats are physiologically similar to human beings.
Rats with prostate cancer, given tomato powder, showed a marked decrease in rate of death compared to rats with prostate cancer that were not given the tomato powder.
So, tomatoes and tomato-based products will prevent prostate cancer in humans.

Here the first premises is a statement of comparison between rats and humans with nothing specific being compared (i.e., there are no attributes given). The example can be and often is formally presented as follows in a simplified manner:

S is similar to T.
S has attribute α.
So, T has attribute α.

This second commonly seen formulation is considerably vaguer than the other one. It merely indicates that the subject terms are similar, without stating how they are similar. This version is nearly impossible to evaluate.

Overall, analogical arguments, however formulated, are quite difficult to assess and perhaps only rarely make for strong arguments. Still, analogical arguments, especially in modelling, are indispensable for scientists—if only because of their heuristic value.

Deductive Arguments

Deductive arguments also have an important role in scientific practice.

A DEDUCTIVE ARGUMENT is an argument in which the premises purport to provide absolute support for the conclusion.

Below lists two forms of deductive argument that are common in science: modus tollens and disjunctive syllogism.

A deductive argument is *valid* when its premises, when assumed true, satisfy the Condition of Relevance and give one no alternative other than to accept the conclusion as true.

A DEDUCTIVELY VALID ARGUMENT is a deductive argument whose premises, when assumed true, do not allow for the falsity of the conclusion.

The right sort of deductive argument, one that is *sound*, has true premises that give one no alternative other than to accept the conclusion as true.

A DEDUCTIVELY SOUND ARGUMENT is a deductive argument whose premises, when actually true, do not allow for the falsity of the conclusion.

Modus Tollens (Denying the Consequent)

MODUS TOLLENS, meaning the denying way, is a valid deductive argument with three statements: a conditional premise, a premise that denies the consequent (then-part) of the conditional premise, and a conclusion that affirms the denial of the antecedent (if-part) of the conditional premise.

This is tremendously important form of deductive argument according to hypothetico-deductivists and falsificationists (see Module 8). It has the following form:

1 If A, then B.
2) Not B.
3) So, not A.

Here A and B can stand for any statement whatsoever, however complex.

Example:

[1]If the bodies of the stars moved in a quantity either of air or of fire diffused throughout the whole (as everyone assumes them to do), the noise which they created would inevitably be tremendous, and this being so, it would reach and shatter things here on earth. Since, then, [2]this obviously does not happen, [3]their motions cannot in any instance be dues either to soul or to external violence. (Aristotle, *On the Heavens*)

Formalized:

If (A) the bodies of the stars moved in a quantity of air or of fire diffused throughout the whole, (B) the noise which they created would inevitably be tremendous and it would reach and shatter things here on earth.
We do not hear or see the effects of such a noise on earth (~B).
So, the bodies of the stars do not move in a quantity of air or of fire diffused throughout the whole (~A).

In conditional sentences (i.e., if-then sentences), the truth of the antecedent (if-part) is said to be sufficient to guarantee the truth of the consequent (then-part). This is, in effect, just what we mean by the if-then relationship in sentences. We mean that the sentence as a whole is true whenever the if-part is true and the then-part is true. For instance, let us take a conditional sentence where antecedent and consequent are semantically linked.

When someone has a dianthus, she also has a flower.
Angela does not have a flower.
So, Angela does not have a dianthus.

Here having a dianthus would be sufficient for having a flower (since a dianthus is by definition a type of flower), but the second premise tells us that Angela does not have a flower, not that Angela has a dianthus. Because a dianthus just is a type of flower, not having a flower is a guarantee of not having a dianthus. So, we must conclude that Angela does not have a dianthus.

Of course, if-then relationships are not always semantic like the example above. They are, for example, often causal ("If you behead a person, you bring about his death") or even ridiculous ("If Dubya gets elected again, then I'll move to Afghanistan"). These, however, do not impact the validity of modus tollens. One may always conclude that any conditional, whose consequent is denied in another premise, has an antecedent that *must* be false.

Disjunctive Syllogism

> **DISJUNCTIVE SYLLOGISM is a valid deductive argument that may be condensed to three statements: a disjunctive premise, a premise that denies all but one of the disjunctive statements, and a conclusion that affirms the one remaining disjunctive statement that has not been denied.**

It has the following form:

1) A, B, or Γ.
2) Not A and not B.
3) So, Γ.

There are no limits to the number of disjunctive parts (here, signified by A, B, and Γ) in the first statement: There can be two or twenty-two. It is only needed that the second statement deny all but one of them for the argument to go through. I illustrate below by reference to the Ptolemaic (earth-centered) and Copernican (sun-centered) systems of the universe (see Module 13).

Example:

> The universe is either (A) a Ptolemaic system or (B) a Copernican system.
> The telescopic observations show that the Ptolemaic account cannot be true (~A).
> So, the Copernican account is true (B).

The chief difficulty with disjunctive syllogisms in science is that they are often unsound, because the first premise is not true. We need only to consider the argument given above. For one, its truth could not be guaranteed in Copernicus' day, because the two systems were not exhaustive. As an example, there was Tycho Brahe's system—a sort of compromise between the Copernican and Ptolemaic models. It had the earth in the middle of the universe, the moon and sun revolving around the earth, and all the other planets revolving around the sun. If the first premise is not exhaustive, then there needs to be some guaranteed that the true alternative is in one of the disjunctive parts. There cannot. It is as if one were to solve a murder mystery conclusively by rounding up five suitable suspects and concluding that the fourth must be the killer by showing that the other four did not commit the crime.

Aristotle's Telic Cosmos

Aristotle's cosmos is a plenum—i.e., it is completely filled with objects that are, in principle, sensible. Sensible objects, for him, comprise matter and form. Form inheres within each sensible object and it gives shape, function, and essence to that object, which, considered otherwise, is merely a mass of matter. Form, then, is the essence or defining characteristic of a thing. To identify something's form is to know its definition, purpose, or function. All entities, both living and non-living, have a particular form and, presumably, all entities of the same kind have the same form (i.e., every penguin has the form of penguin). Matter, easily enough, is an entity's material composition. Matter determines the possibilities of something's use, for the kind and quantity of matter determine just what forms it can acquire. From a block of marble, for instance, one cannot make a soft sphere or a sphere that is larger than the block. Things of the same form appear and are different only insofar as they have different matter. So, as form is a principle of taxonomic identification, matter is a principle of individuation.

Aristotle's cosmos is a teleological system with two main realms (see Figure 3.1). The dividing line of these two realms is the sphere of the moon: All things above the sphere of the moon are perfect, divine, and unchanging; all things below this sphere are generated, destroyed, and changeable. The physical principles regulating both realms, taken from Aristotle's *Physics* and *Metaphysics*, I summarize below:

1. Nature=df: Nature is a principle of change (i.e., motion) or remaining unchanged (i.e., unmoved).
2. Principles of Motion: PM_1: Motion cannot come to be or perish; PM_2: An everlasting motion is initiated and sustained by an everlasting moved; PM_3: A single motion is initiated by a single mover.
3. Principle of Potentiality: What has a potentiality need not actualize it.
4. Principle of Determinism: Nothing is moved at random.
5. Principle of Change: Whatever is moved can be otherwise.

Below the moon's sphere, things comprise the material elements—fire, air, water, and earth—each of which has an active (hot or cold) and passive (dry or wet) capacity (Gr., *dynamis*) and has its own proper sphere to which it naturally tends. Fire (hot and dry), moves by nature away from the center of the cosmos and to the outermost sphere of the sublunary realm—the sphere of the moon. Air (hot and wet) also moves away from the center and finds its proper place just under the sphere of fire, as air is light, but not so light as fire. Earth (cold and dry) is the heaviest element and moves toward the center of the cosmos as do all heavy things. Water (cold and wet) is also heavy and tends toward the center. Being lighter than earth, its proper place is just above the earth and below the sphere of air. So the sphere of the moon is a limit for the motion of light elements, while the center of the cosmos is a limit to natural downward motion. What prevents the elements from settling in their proper spheres in homogeneous masses is the circular motion of the sun and, to a lesser extent, that of the moon. These motions mix the elements and enable them to form homogeneous and heterogeneous masses as parts of living things.

Above and including the sphere of the moon, there are some 55 concentric spheres that account for the stars and planets, as well as the initiating and sustaining cause of all motion—the prime mover. Each of these spheres is a divine body. Outside of this last sphere exists the prime mover—the most divine substance. Being pure actuality and pure form, the prime mover initiates and maintains the motion of the spheres through constant and self-contemplation. The superlunary realm outside is filled with an unchanging fifth material element that is endemic—*aether*. All motion in here is unending and circular.

In all, Aristotle's cosmological account aims to explain how the cosmos itself possesses the good. He offers, by analogy, the goodness in order in an army. The good of this order is some actualized potentiality of the men that is realized not because of the men themselves, but because of the general in charge of the men. Similarly, the good of the cosmos is caused by the most divine first mover as governor of all order.

> All things—fishes, birds, and plants—are joined in some order, but not all in the same way. Nor are they unrelated to each other, but they have some relation; for all things are joined in some order in relation to one thing. (It is like a household, where the free members are least of all at liberty to do what they like, and all or most of what they do is ordered, whereas only a little of what slaves and beasts do promotes the common [good], and mostly they do what they like.) For the nature of each sort of thing is such a principle [that aims at the good of the whole]. I mean, for instance, that everything necessarily is eventually dissolved, and in this way there are other things in which everything shares for [the good of the whole]....

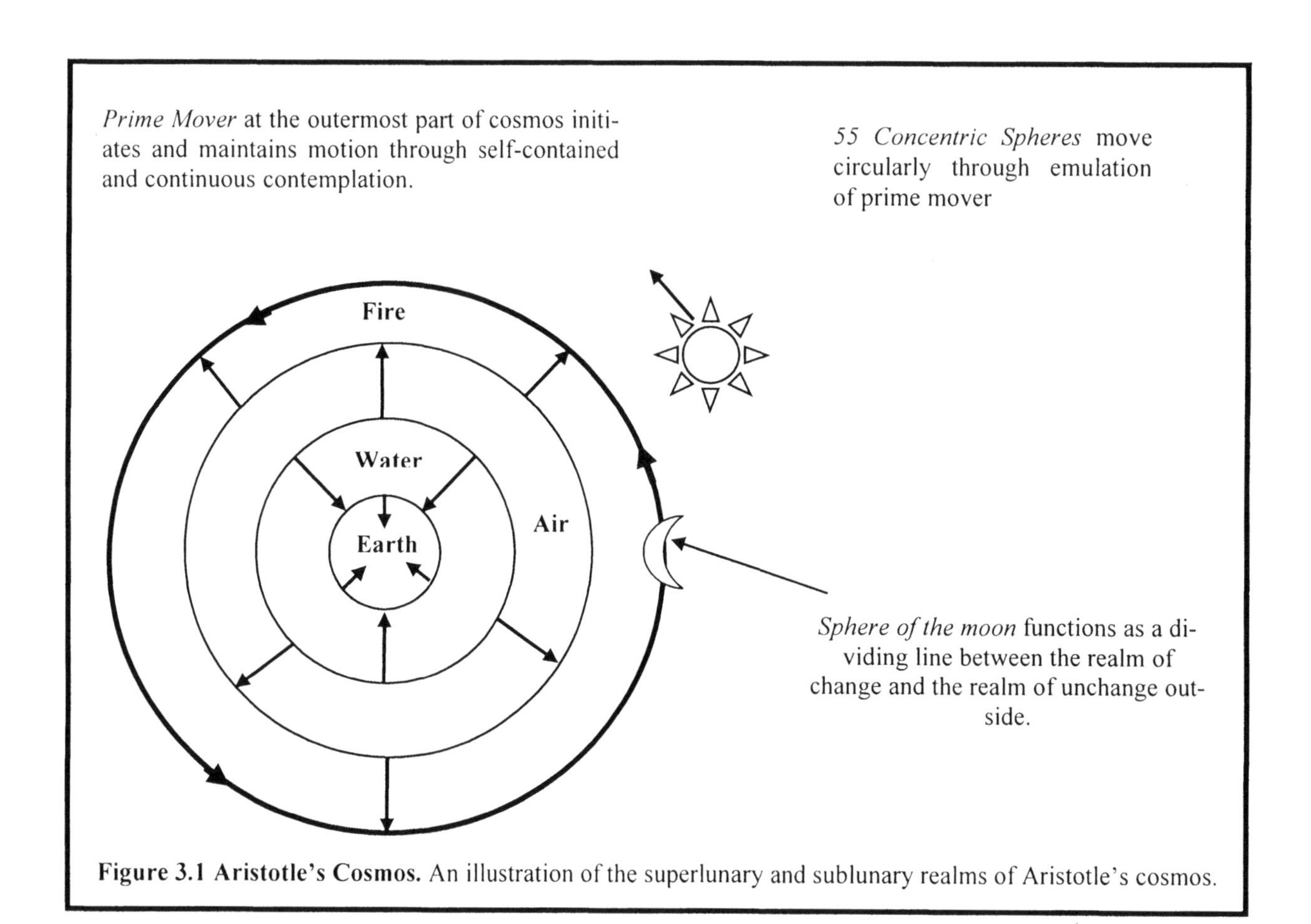

Figure 3.1 Aristotle's Cosmos. An illustration of the superlunary and sublunary realms of Aristotle's cosmos.

Key Terms

inductive argument
cogent argument
inductive generalization
modus tollens
hasty generalization
principle of sufficient variation
random sample
Aristotle's cosmos
deductive argument
sound argument
analogical argument
disjunctive syllogism
biased sample
principle of sufficient size
Aristotle on knowledge

Text Questions

* What distinguishes an inductive argument from one that is deductive?
* What makes an inductive argument cogent? What makes a deductive argument sound?
* What are the two types of fallacy one can have with inductive generalizations?
* Why should one be suspicious of analogical arguments?
* How might modus tollens or disjunctive syllogism be used by scientists to decide between competing hypotheses for some phenomenon under consideration?

Text-Box Questions

* How does Aristotle think we can derive knowledge from sensory experiences?
* Aristotle's telic cosmos was the predominating view of the cosmos for some 2000 years. Why might his telic world-view be called a proper-place cosmology?

Exercises

Identify each of the following arguments as an instance of inductive generalization, argument from analogy, modus tollens, or disjunctive syllogism.

1. If the origin of man had been wholly different from that of all other animals, these various appearances [rudimentary and useless structures] would be mere empty deceptions; but such an admission is incredible. These appearances, on the other hand, are intelligible, at least to a large extent, if man is the co-descendant with other mammals of some unknown and lower form. Darwin, *The Descent of Man*

Argument type: ____________________

2. While a flowing stream is corrupted either with difficulty or not at all, confined water is easily corrupted; thus, with a river of eloquence, the censures of refutation are removed, while the poverty of confined reason does not defend itself easily. Cicero, *Tusculan Orations*

Argument type: ____________________

3. Galileo, measuring the rate of fall of balls down grooved channels of differently inclined planes, argues:

This operation being precisely established, we made the same ball descend only one-quarter the length of this channel, and the time of its descent being measured, this was found to be precisely one-half the other. Next making the experiment for other lengths, examining now the time for the whole length [in comparison] with the time of one-half, or with that of two-thirds, or of three-quarters, and finally with any other division, by experiments repeated a full hundred times, the spaces were always found to be to one another as the squares of the times. And this [held] for all inclinations of the plane: that is, of the channel in which the ball was made to descend, where we observed also that the times of descent for diverse inclinations maintained among themselves accurately that ratio that we shall find later assigned and demonstrated by our Author. Galileo Galilei, *Dialogue Concerning Two Chief World Systems*

Argument type: ____________________

4. If the stars moved of themselves they would naturally perform their own proper motions; but we see that they do not perform these motions; therefore they cannot move of themselves. Aris-

totle, *On the Heavens*

Argument type: ____________________

5. If the Son of God was obliged to sacrifice his life to redeem mankind from original sin, then by the law of talion, the requital of like by like (i.e., an eye for an eye), that sin must have been a killing, a murder. Nothing else could call for the sacrifice of a life for its expiation. And the original since was an offence against God the Father, the primal crime of mankind must have been a parricide, the killing of the primal father of the primitive horde, whose mnemic image was later transfigured into a deity. Sigmund Freud, "Thoughts for the Time on War and Death"

Argument type: ____________________

6. From Darwin's own notebooks, prior to articulating his theory of evolution:

The affinities of all the beings of the same class have sometimes been represented by a great tree. I believe this simile largely speaks the truth. The green and budding twigs may represent existing species; and those produced during each former year may represent the long succession of extinct species. At each period of growth all the growing twigs have tried to branch out on all sides, and to overtop and kill the surrounding twigs and branches, in the same manner as species and groups of species have tried to overmaster other species in the great battle for life.... As buds give rise by growth to fresh buds, and these, if vigorous, branch out and overtop on all sides many a feebler branch, so by generation I believe it has been with the great Tree of Life, which fills with its dead and broken branches the crust of the earth, and covers the surface with its ever branching and beautiful ramifications. Charles Darwin, *Origin of Species*

Argument type: ____________________

7. The states of the soul by which we always grasp the truth and never make mistakes, about what can or cannot be otherwise, are scientific knowledge, practical wisdom, wisdom, and intuition. But none of the first three—practical wisdom, scientific knowledge, wisdom, is possible about origins. The remaining possibility, then, is that we have intuition about origins. Aristotle, *Nicomachean Ethics*

Argument type: ____________________

8. A theory-less position is possible only if there are no theories of evidence. But there are theories of evidence. Therefore, a theory-less position is impossible. Henry W. Johnstone, Jr., "The Law of Noncontradiction", *Logique et Analyse*

Argument type: ____________________

9. Lastly, the findings of my analysis are in a position to speak for themselves [i.e., that hysteria is brought about by childhood sexual abuse]. In all eighteen cases [cases of pure hysteria and of

hysteria combined with obsessions, and comprising six men and twelve women] I have, as I have said, come to learn of sexual experiences of this kind in childhood. Sigmund Freud, "The Aetiology of Hysteria"

Argument type: ______________________

10. [S]uppose I had found a watch upon the ground, and it should be inquired how the watch happened to be in that place, I should hardly think of the answer which I had before given—that, for anything I know, the watch might have always been there. Yet why should not this answer serve for the watch as well as for the stone? ...For this reason, and for no other, viz., that, when we come to inspect the watch we perceive (what we could not discover in the stone) that its several parts are famed put together for a purpose.... This mechanism being observed,...the inference, we think, is inevitable, that the watch must have had a maker...who comprehended its construction, and designed its use. William Paley, *Natural Theology*

Argument type: ______________________

Module 4
Critical Reasoning & Science

> "By always thinking unto them (do I make discoveries). I keep the subject constantly before me and wait till the first dawnings open little by little into the full light". Sir Isaac Newton, *His Life and His Work*

What Is a Hypothesis?

WILMA JO JONES WALKS INTO HER DOCTOR'S OFFICE and tells her doctor that she has a growth beneath her jawbone. The doctor sits her down to examine the growth. His initial concern is that the growth is cancerous. He looks inside of her mouth and examines the tissue around the growth. He asks her whether it is painful and about the length of time that she has had it. Wilma Jo replies that she may have first noticed it several years ago, as she vaguely recalls one night noticing a small growth in the same area, but then she talked herself out of any worry about it. The doctor listens to her, examines it again, and this time even squeezes the growth ever so slightly and then squeezes it again firmly.

After deliberation, he tells her that he cannot definitely rule out cancer, yet every indication points to an inexplicable growth of fatty tissue—a lypoma. He says the growth is too soft and too malleable for cancer. Moreover, if she had first noticed it many years ago and it was cancerous, it likely would have spread to other areas of her body, which show no indication of tissue degeneration. She is calmed, but the doctor says that he wants to be sure, so he wants to schedule for a biopsy of the growth to examine the tissue and to be certain that it is not cancerous. Weeks later, the results of the biopsy show no indication of cancer. Wilma Jo is relieved.

Wilma Jo's doctor here has been engaged in scientific activity. A patient enters with a *problem*. He listens to her story and frames possible reasons for the growth—that is, tentative *hypotheses*—that are consistent with her account. He also listens for details that would suggest that some of the hypotheses he has formed are improbable. That the growth is not painful and that it has likely been around for several years offer evidence that it is not cancerous. Next, he examines her closely to see which of the possibilities, if any, can be ruled out by a brief physical inspection. He notices the softness of the tissue, which is more in keeping with fatty, not cancerous, tissue. He looks to see whether other parts of her body, like the mouth, have growths and finds none. Later on, the results of the biopsy, which have disclosed no cancerous tissue, strongly confirm his suggestion that it is fatty tissue and not cancer.

Let us summarize below what the doctor has done for Wilma Jo. The incident begins with a particular problem. Listening to the patient's account, the doctor formulates initial hypotheses—here fatty tissue or cancer—in an effort to explain the phenomenon to be investigated. These initial hypotheses, in turn, suggest certain methods of how to proceed with his examination. He then looks for evidence in favor of the one hypothesis and, conversely, any evidence that rules out the other hypothesis. This, in large part, is the method of science. Summed below:

PROBLEM PHASE: *Science begins with a problem that prompts inquiry.*

Hypothesis Formulation: *This problem invites speculation in the form of hypotheses in an effort to solve the problem. It is important here that the scientist considers every reasonable hypothesis. These hypotheses may be mere conjectures or they may be guided inductively, by analogies or past experience.*

Scientific Prediction: *These hypotheses, in turn, have predictive consequences that offer tests that either confirm or disconfirm them.*

Data Gathering: *The practitioner does whatever testing he can in an effort to gain as much information as possible to make the best-informed decision.*

Evaluation: *Last, the scientist evaluates the data and gives a verdict, so to speak, as to how the data relate to each hypothesis under consideration.*

What this episode shows is that two different types of thinking go into any attempt to solve a scientific problem. To borrow from timely metaphors, these two types of thinking can be called *thinking outside-the-box* and *thinking within-the-box*.

Plato's "Likely Story"

Plato's (427-347 B.C.) thoughts on sameness and difference—the oneness and inalterability of reality, but the flux and plurality of visible things—are spelled out in *Republic* VI-VII. Here Plato takes a fresh look at Parmenidean metaphysics, not by challenging any of Parmenides' assumptions on what is real, but through offering an explanation of reality, consistent with those principles, that does not entirely explain away visible phenomena. The results are his doctrine of the divided-line and his theory of forms.

In effect, *Republic* establishes a rationalist compromise between Heraclitean (all things are in flux), Pythagorean (all things are numbers), and Parmenidean (reality is one, unchanging) metaphysical principles. The difficulty is that this compromise is not much of an explanation of visible objects. Plato's reason for neglecting the visible world in *Republic* (and elsewhere) is that visible objects, being ever-changing, are not proper objects of knowledge, but mere objects of opinion. Plato, like Parmenides, draws a sharp line between objects of intellect and objects of sensation—and only the former are worthy of true philosophical investigation.

Nevertheless, Plato does grapple with a causal account of the world of visible things in a work entitled *Timaeus*. In keeping with the low regard Plato has of physical reality, he tells us, through the mouth of Timaeus, that he can do no more that provide a "likely story" when he tackles issues such as cosmic teleology, acoustics, harmonics, human physiology, and human psychopathology. Though these particular accounts are likely not his own, the overall teleological vision of Timaeus is uniquely Platonic.

The cosmos had its origin due to a divine craftsman, who employed two productive forces, Reason and Necessity, to put together the matter available for cosmic construction in the best possible manner. At the cosmic level, the uniform and regular motions of the westward-moving sphere of the heavens and the eastward-moving circle of the ecliptic result in a helical movement of the Sun between the tropics. At the subcosmic level, the materials out of which all things are constructed—fire, air, water, and earth—are explained by four even more basic geometric solids: the pyramid, the octahedron, the icosahedron, and the cube. These, in turn, are reduced to two basic triangles: the half-square and the half-equilateral. All of this occurs in the great "Receptacle of Becoming": space. Thus, in strict geometric fashion, the divine craftsman arranged the cosmos.

Thinking outside the Box

In the case of Wilma Jo, we notice that her doctor does whatever he can to gain preliminary information about the tissue growth on her neck. His many years in medical school and vast

Law_1: The orbit of each planet is the shape of an ellipsis with the sun located at one focus (*The New Astronomy*, 1609).

Law_2: In any equal interval of time, a line from a planet to the sun will sweep out equal areas (i.e., ASB = CSD = ESF, where the time from A to B = the time from C to D = the time from E to F) (*The New Astronomy*, 1609).

Law_3 (See Below): The ratio of the square of the time (T) of revolution around the Sun to the cubes of their mean distances (D) from the Sun for a given planet (including Earth) is equal to the same ratio for any other planet, or $D^3/T^2 = k$ (*The Harmony of the World*, 1916).

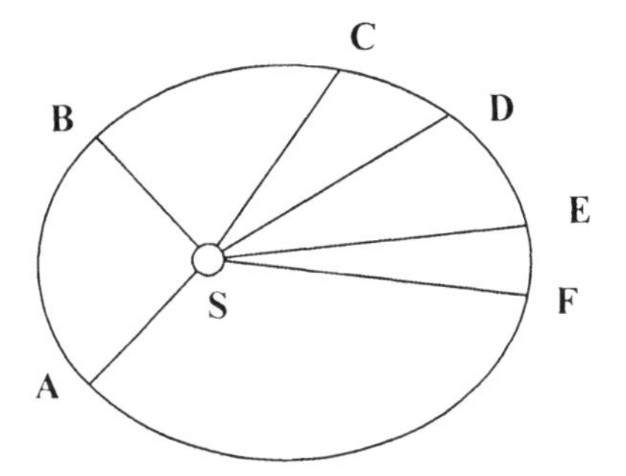

Here, area ASB = area CSD = area ESF, in accordance with Law_2.

Figure 4.1 Kepler's Laws of Planetary Motion.

amount of clinical experience help him to formulate two hypotheses related to the problem. His reasoning here seems mostly limited by medical theories about such things as well as his own experiences. Still, however, each and every case is unique. Theory has limited application and experience too is circumscribed by what one has experienced. These seem helpful, even indispensable, guides for his making an informed decision, but they are not infallible guides. Yet at some point, he does make a decision and this decision, fallible as it inevitably must be, is the best he can do at this point in time. With his decision, the physician has made a leap, as it were. He states with resolve that the tissue is fatty and not cancerous. Wilma Jo is relieved, as she has complete trust in her doctor. What other option does she really have?

Of course, Wilma Jo might not have been so relieved had her doctor behaved otherwise. Imagine that her doctor suggested immediate surgery and, when asked why by Wilma Jo, he responded, "I don't have a clue, when it comes to things like abnormal tissue growths. I only know that I like to cut out things that don't look so good to me". One objection at this point is that this example is ludicrous. Scientists just do not act this way. The response to this objection is quite simple: Some scientists often do act this way. Sir Isaac Newton, for instance, postulated that *gravity was an attractive force between masses that acted in inverse proportion to the distances of the masses* ($G = M_1M_2/d^2$). Though the theory was an overwhelming success for over 200 years, he could never come to grips with how one body could act on another from a distance, as everything known of bodily interaction at the time demanded contiguity of the bodies interacting. When asked how such a force could act at a distance, Newton famously replied, "Hypotheses non fingo" (literally, "I frame no hypotheses" or, more appropriately, "I don't have a clue"). Newton, however, did not see himself as engaging in constructing hypotheses, as he judged this to be outside of proper "experimental philosophy".

It is well known that Johannes Kepler (1571-1630), before Newton, was a bold, and often incautious, theorizer about the construction of the heavens. He sometimes drew misguided inspiration from the Greek philosopher and occultist, Pythagoras, who lived well over 1500 years prior to him (see Module 2). Nonetheless, Kepler's lack of conservatism enabled him, over time, to come up with the correct model of the orbits of planets (see Figure 4.1, above).

For centuries, it was assumed, following the lead of Aristotle in the fourth century B.C., that whatever explanation there might be of celestial motions, this explanation must be couched in terms of these bodies traveling with *uniform and circular motion*. The reason Aristotle gave was roughly that only uniform, circular motion was befitting such divine-like objects. Now Kepler had access to relatively accurate measurements of planetary positions, which he inherited from his mentor Tycho Brahe, the greatest observational astronomers of his day. Kepler tried in vain to discover models that fit the observations. Success finally came when he gave up the notion that planets must travel circularly and with uniform motion. Through brilliant persistence with this problem and an unshakable belief that celestial phenomena are mathematically explicable, he eventually found that if he began with the assumption that planets travel in elliptical, not circular, orbits, he could develop planetary models that fit the phenomena quite elegantly.

Let us look also at Sigmund Freud's Oedipus complex. Freud has been and continues to be often maligned as a psychologist, because he was impacted by Greek mythology in coming up with his Oedipus complex. In Sophocles' *Oedipus Rex*, it is foretold by the oracle that Oedipus would kill his father, King Laius of Thebes, and marry his mother, Queen Iocasta. The story by Sophocles goes on to play out this theme. Drawing from this myth and his own sexual feelings for his father's second wife (who was younger than Sigmund), Freud devised a theory that there is Oedipal sexual tension in every case of normal adolescent sexual development. According to Freud, having been nurtured by his mother, a boy soon sees himself as a rival of his father for his mother's affection. This rivalry is fundamentally sexual. Of course, what he feels for his father is what any rival should feel—hatred. His natural impulse, in order to win his mother's affection, is to kill his father. The situation is resolved, insofar as it can be resolved, by a literal castration threat to the boy. This threat poses so great a menace to the boy that his sexual longing for his mother becomes repressed—more specifically, Freud says, expunged. So too is his hatred of his father. With repression, the boy's conscience or super-ego develops. He unconsciously renounces his lust for his mother and his hatred of his father. He then comes to identify with his father and, not being able to overcome fully his feelings for his mother, seeks to marry someone much like her.

There are two important points to consider about the Oedipus complex. First, Freud unmistakably states that the Oedipus complex is universal in scope—that is, it applies to all boys everywhere. Thus, no attempt to resuscitate Freudian psychoanalysis by positing that the Oedipus complex applies only in certain family complexes of certain cultures will do. Second, the Oedipus complex is the very cornerstone of Freudian psychoanalysis. Freud was not shy about this. In certain works (see Module 25), he went so far as to root the Oedipus complex biologically in stating that the complex arose through the actions, in early tribal history of humans, of a group of "sons", who banded together to kill their "father" or leader of the tribe. It was an act so despicable that it was completely forgotten. Hence, Freud attempts to give a historical origin to the unconscious and the super-ego.

Whatever feelings one has about Freud or the Oedipus complex, saying that the theory is untenable because Freud drew inspiration from Greek mythology is not proof that the theory is false. Scientists draw inspiration from the strangest sources. Friedrich Kekule came up with his model of benzene as a hexagonal ring from a dream, in which he saw a snake biting its own tail.

Freud on the Oedipus Complex

In 1923, Sigmund Freud wrote *The Ego and the Id* and here posited his Dynamic Model of the human psyche—comprising the id, the ego, and the super-ego.

The *id*, he said, is our psychic apparatus at birth and filled with phylogenic memories (i.e., memories of past deeds by prior members of our species). If functions according to the pleasure principle, which is essentially of a sexual nature. Failure to discharge the sexual impulses of the id results in neurosis.

The second psychic agency to develop is the *ego*. This forms when a child comes to distinguish between his self and other objects. Forming thus, it is bound by considerations of reality. It functions principally to delay the gratification the id seeks on account of reality. As the major function of the ego is to effect compromises between id impulses and the demands of reality and as these are roughly equivalent to defenses, ego operations are commonly called "defense mechanisms".

Finally, by the age of five or six, the ego-ideal or *super-ego* develops. This is the moral agency of a child that functions independently of the ego to permit and prohibit actions, in accordance with conscience.

Just how for Freud does the super-ego come about? Here is where the Oedipus complex comes in.

A young boy's first object of affection is his mother—specifically her breast. At the same time, he identifies with his father. Through caring for the boy's body, his mother shortly becomes his seducer. However long he is weaned, it is always too little and too short. He comes to see himself as his mother's lover and is proud to show his organ to her and share her bed when his father is gone. Since his father has a larger penis, his father is a genuine threat for his mother's affection. He becomes a rival for his mother's affection. The boy becomes ambivalent toward his father: He both loves him as father, but he also hates him as a rival. This ambivalence is the key to the development of the ego-ideal.

As the mother becomes aware of the boy's intense feelings for her, she subsequently forbids him to handle his organ. She threatens castration, usually through a verbal threat by his father. This threat then becomes the most severe trauma in the boy's life. He must either identify with his mother or renounce his love for his mother and identify with his father. Which alternative the boy follows is generally decided quantitatively: Each person is constitutionally bisexual and has certain quantities of masculine and feminine psycho-sexual impulses.

In most cases, masculine sexual energy determines that he will renounces his mother and his own masculinity so as to preserve his penis. He gives up his mother as a love object and his feelings become completely forgotten or repressed. He becomes passive toward his father. His hatred toward his father turns into a more complete identification with him. This massive repression, founded on ambivalence toward his parents, results in the formation of the ego ideal or the super-ego.

> The broad general outcome of the sexual phase dominated by the Oedipus complex may, therefore, be taken to be the forming of a precipitate in the ego, consisting of these two identifications in some way united with each other. This modification of the ego retains its special position; it confronts the other contents of the ego as an ego ideal or super-ego.

Thus, the super-ego is an enormous defense effort directed toward a boy's Oedipal impulses and his sense of guilt. This Freud takes to be the biological component of the super-ego.

This, of course, is a rather queer origin of a scientific hypothesis. Still the model turned out to be correct. The moral of the story is worth repeating: *Scientist draw inspiration from the strangest sources*. Ultimately, what is important is not so much how a scientist comes to adopt a particular hypothesis, but rather whether or not the hypothesis is true.

What all of this is meant to show is that not all scientific thinking is or ought to be formally constrained. Sometimes, fruitful imagination is a much more valuable asset in framing a hypothesis than are sound thinking skills. As Einstein once said, "Imagination is more important than knowledge".

Thinking outside of the box is a matter of problem-solving outside of set, abstract parameters.

Thinking within the Box

There is, of course, when considering Wilma Jo's case above, a great deal of thinking within the box that goes on as well. The doctor is not, of course, at liberty to draw just any arbitrary conclusions from the evidence before him. Normative rules of thinking have a role. Moreover, his vast experience and education will impose strict limits on the preliminary hypotheses he puts forth. One can imagine, however, that experience and education are not always good things. There will always be cases to which experience and training do not apply, as when experience and training do more to hinder than to help.

Thus, having put forth his preliminary hypotheses, Wilma Jo's doctor will go on to test these hypotheses. If the lump is cancerous, it should feel relatively hard and be somewhat difficult to move. Additionally, if she has had the cancer for several years, there should be cancerous growths in other parts of the body. If however it is fatty tissue, it will be softer and more inclined to move around and no cancerous spots should be found in other parts of her body. And so on. These are the types of tests any good physician, with similar training and education and correct reasoning skills, will be constrained to do.

This last claim may seem quite obvious to most people. The reason for this is that there are right criteria of thought that apply to all people everywhere at any given time. What would Wilma Jo think of her doctor if he explained his reasoning thus: "Fatty tissue is soft, while cancer, in general, is firm. This tissue is definitely soft, so this tissue is cancerous". Her immediate reply would be something like: "I think you meant to say that this tissue is *not* cancerous". If he stubbornly insists that he said precisely what he meant to say, then Wilma Jo would be justified, I suppose, in thinking that her doctor is not completely sane. *Anyone* reasoning like her doctor here is reasoning incorrectly. This is reasoning within the box. This is critical thinking as logical thinking—its normative dimension.

Thinking within the box is a matter of problem-solving within the confines of set, abstract parameters of evaluation.

Three Types of Hypotheses

Before ending this section, I would like to introduce the three types of hypothesis, fully examined in later modules: universal, statistical, and causal hypotheses.

Universal hypotheses are so-called because they have universal scope. The following claims are each universal:

All eagles are predatory.
Dreams are generated by unconscious wishes.
In natural processes, the entropy of the world always increases.

The middle claim is not explicitly, but implicitly universal, as the quantifier "all" is implicitly understood to be linked with "dreams".

Statistical hypotheses, in contrast, involve propositions that are statistical in nature.

> A female born in the United States between 2000 and 2005 has an 86.4% chance of living to 65 years of age.
> Lionesses conduct 90% of the hunting done by all lions.
> There is a good chance that the sun will become a white dwarf star within the next six billion years.

The last claim is statistical, but vague. It states that there is more reason than not to believe the sun will become a white dwarf in the next six billion years.

Last, there are *causal hypotheses*.

> Passing light near a massive object causes it to bend.
> Oxygen is the cause of flame.

To get clear on causal hypotheses, it is critical to have a clearer understanding of causation—the subject of Section Three, Modules 7-9. It will suffice for now to say that causal hypotheses do more than state that certain variables are correlated. Ascription of causal force, in common speaking, is to say that one event completely determines another (example one, above) or that one event cannot occur without another (example two, above).

KEY TERMS

within-the-box thinking
scientific hypothesis
statistical hypotheses
Newton's law of gravity
Freud's dynamic model
outside-the-box thinking
universal hypotheses
causal hypotheses
Kepler's laws of planetary motion
Freud's Oedipus complex

TEXT QUESTIONS

* Roughly speaking, what is the method of science? How does outside-the-box thinking come into play? How does within-the-box thinking come into play?
* What are the three types of hypotheses?

TEXT-BOX QUESTION

What is the Oedipus complex for Freud? Why is it unfair to say that the complex cannot be true merely because the inspiration behind it was most unusual?

PART II

PHILOSOPHY OF SCIENCE WITH AN INVESTIGATIVE EYE

SECTION TWO

Science & Truth

Module 5
Theories of Truth

"Perhaps the good man differs from others most is seeing the truth in each class of things". Aristotle, *Nicomachean Ethics*

AS THE NEXT FOUR MODULES WILL SHOW, many of the problems that plague philosophers of science are problems generated by there being no one commonly accepted theory of truth. So before turning to these difficulties, it is necessary to sketch out some of the most commonly held theories of truth as they relate to philosophy of science.

Correspondence Theory of Truth

Aristotle, about 2400 years ago, wrote, "To say of what is that it is not, or of what is not that it is, is false, while to say of what is that it is, and of what is not that it is not, is true"[1] This is the earliest unambiguous account of the correspondence theory of truth.

> **According to the CORRESPONDENCE THEORY OF TRUTH, a statement is true if what it asserts of the world is in fact true of the world.**

Truth, according to the correspondence view of truth, is a relatively straightforward mapping of words on to the world. I use the word "mapping" intentionally, as maps are constructed to represent structurally what it is they are intended to represent. It is the same with propositions. Consider, for instance, two sentences:

S_1: Earth is the third planet from the sun in our solar system.
S_2: Jupiter is the third planet from the sun in our solar system.

According to the correspondence theory of truth, S_1 is true because Earth is in fact the third planet from our sun, while S_2 is false because Jupiter is in fact the fifth planet from the sun. While the first "maps on" to the world, the second fails to do so.

This account of truth seems quite sensible at first blush. Yet there are problems. The main difficulty with the correspondence theory of truth is that sentences do not seem to map on to reality so straightforwardly. To put it more precisely, most philosophers concede that we have no way of ascertaining the "facts" of the world independently of certain conceptions about the way the world is. For example, one can easily see how the same sentence can mean two different things to two different people. Consider the sentence, "Snow is cold". This seems unquestionably true to Ramsey, who has lived his whole life in Ethiopia and had his first experience of snow on a trip to Germany. To Natasha, who has spent the lion's share of her life in Northern Siberia, the sentence may be vague or senseless, as she knows that Siberian air makes for different types and temperatures of snow, some of which she might not consider to be cold at all.

Again, consider Wentworth, who believes that whenever someone grasps some object that object is, at that very time it is being grasped, a part of the very person who grasps it and no longer an independently existing object. Now Petrunia comes before Wentworth with two books in her hands. She places one on a desk right in front of her and continues to hold the other in her hands. While doing so, she asks Wentworth, "How many books do you see before you?" To Petrunia's amazement, Wentworth answers, "One", as the book in Petrunia's hand is not seen to be an independently existing object, but rather a part of Petrunia's very person. What these illustrations are meant to show is that our preconceptions of the world determine what it is we call "facts".

A reply to this objection, at least at the level of science, is that each science has a language of its own that is universally accessible to any and all who wish to study that science. The language of particle physics is principally mathematical and does not vary much at all from culture to culture. Terms in chemistry are fixed by chemists—through observation, experiment, and neologism—over time. Even psychotherapeutic practice, through its continually revised and updated DSM manual (*Diagnostic and Statistical Manual of Mental Disorders*), is working slowly toward a common language, though agreement here is not always reality-driven.

Coherence Theory of Truth

The next theory is the coherence theory of truth—a popular alternative to the correspondence theory.

> **According to the COHERENCE THEORY OF TRUTH, a statement is true if it is logically consistent with other statements within a propositional system that are accepted as true, that mutually support each other, and that can offer together a relatively complete picture of reality.**

With the coherence theory, the key is internal consistency. No statement can be judged true or false in isolation from a commonly accepted and internally consistent body of other statements. A scientific theory, for example, is one such system. On a larger scale, the commonly accepted body of scientific knowledge at a particular time is another. According to the coherence view, a scientific theory is likely to be true if it is internally consistent and it agrees with other commonly held scientific views about the way the world works. Unlike correspondence, truth here is not a matter of fitting "facts" to an independently existing reality, but of squaring things believed to be true of the world with other beliefs about the way the world works.

The coherence account of truth has had and continues to have a wide following, especially in philosophy of science. Of its defects, I mention four.

First, consider the following two sets of propositions concerning how new species are formed over time: that of Darwin and that of Eldridge and Gould.

Darwin on Speciation over Time:

1. All living organisms have evolved over the course of hundreds of thousands of years from simpler organisms.

2. The change from one species into another occurs gradually but persistently over hundreds of thousands of years.

Eldridge and Gould on Speciation over Time:
1. All living organisms have evolved over the course of hundreds of thousands of years from simpler organisms.
2. The change from one species into another occurs in fits during relatively small geological intervals of time (as little as 5,000 to 50,000 years, according to Eldridge) that are sandwiched between lengthy periods of equilibrium called stasis.

Proposition one is common to both accounts, but proposition two of the first account is inconsistent with proposition two of the second. This makes the two positions inconsistent. Yet both sets of propositions are internally consistent. Which account is true? When coherence *alone* is the criterion of truth and there is nothing else believed to be true about the way the world works to decide between the two views, then there is not any reason to prefer one account over the other, as both are consistent. So, by the yardstick of consistency, both must be true, and this, of course, is absurd, since the two accounts are inconsistent with each other.

Second, consistency by itself provides no assurance that a set of propositions, however large, is grounded in reality. One can, for instance, construct a set of internally consistent propositions about fictitious entities—say, centaurs (P_C)—and then deduce an indefinitely large number of consequences from P_C. The coherence of P_C clearly would have no link with reality, yet by the yardstick of consistency, one is in no position to state that any of these claims is false. One could, of course, appeal to P_C's inconsistency with other systems that are commonly held to be true and consistent with each other, but if these also do not map on to reality, we are in no firm position to reject P_C.

Third, if truth is merely a matter of coherence, then a coherent set of propositions, like P_C, is one in which all of its propositions must be true *by definition*. At the very least, this seems to be putting the cart before the horse. It is astonishingly improbable that, in any very large, internally consistent system of propositions believed to be true about the way the world works, none of them is false.

Finally, coherence presupposes a prior theory of truth. "Coherence" is itself defined, at least in part, as consistency of member propositions and any two propositions are "consistent" when "both can be true at the same time". Judging that any two propositions can be true seems at the same time is the problem. How does one do that? That seems to presuppose an alternative means of measuring their truth and coherence, as a theory of truth, forbids just that.

Relativistic Theory of Truth

The ancient Greek philosopher Protagoras is claimed to have said, "Man is the measure of all things: of things that are, that they are, and of things that are not, that they are not".[2] This was perhaps the earliest articulation of relativism, commonly known today as Protagorean relativism.

According to PROTAGOREAN RELATIVISM, there can be no independent criteria for truth outside of a person's own experiences or judgments, and these experiences and judgments vary from person to person.

What one person judges to be true is true for him; what another person judges to be true for her is true for her. There is no third-person or god's-eye perspective to judge between the two and there is no sense in one person trying to convince another of the falsity of the other's views.

Protagorean relativism, it is not hard to see, is a non-starter. Adopting such a view, there can be no meaningful sense of science, as science requires some common ground for its basis—some minimal level of agreement between scientific practitioners.

Another version of relativism, cultural relativism, is more hospitable to a scientific attitude.

According to CULTURAL RELATIVISM, what is true can vary from society to society or can change within a particular society over time.

In other words, truth is dictated by one's cultural group at a particular time—whether by appeal to consensus within the group or by adoption of the judgments of any representative subgroup within that society that is empowered to make such judgments—not any appeal to an independently existing reality.

Problems with cultural relativism are numerous. Take for instance a particular culture that strives for Aryan supremacy on the utilitarian (i.e., greatest utility for the greatest number) grounds that creating a super race of humans would eventually lead to lack of conflict, global peace, and unsurpassed intellectual and scientific advances. Their means to accomplish these aims is by striving to exterminate all non-Aryans—short-term evil for the promise of long-term and sustained prosperity and peace. Given cultural relativism, another culture with radically different values has no basis for disagreeing with these eugenic utilitarians. To disagree in any substantive sense, there either needs to be some universally agreed-upon canons of judgment or an external, non-subjective means of arbitrating between different points of view of the way the world is. Moreover, take the very statement of cultural relativism that is given above. Let us call it S_{CR}. As uttered by cultural relativists in a particular culture, S_{CR} is merely relative to that culture and that particular time in which it is uttered. If truth is culturally relative, no one from another culture is bound to accept S_{CR}. Yet this does not seem to be what cultural relativists want, for they are making a statement that is applicable to, and true of, all cultures everywhere. If so, then it itself ceases to be a statement that is culturally relative, which is just what the statement forbids.

Protagoras (c. 490-c. 420 B.C.), the Sophist

Protagoras of Abdera was a Greek sophist. In general, sophists were teachers, who traveled from city to city and offered their services to willing clients for a fee. Protagoras taught in Athens and was a friend to the Athenian statesman Pericles. Along with his statement of radical relativism, he is also known for his boast that he can make a weaker argument appear stronger and for his agnosticism concerning the existence of the gods. His relativism and agnosticism are likely the result of the distinction in his day between nature (Gr. *physis*) and custom (Gr. *nomos*), and his belief that humans are in no position to be certain about anything concerning the former.

Pragmatic Theory of Truth

The final view to consider, and one that is very attractive to philosophers of science today, is the pragmatic theory of truth.

> **According to the PRAGMATIC THEORY OF TRUTH, a statement is true if it works—that is, if it yields satisfactory results.**

Of course, everything here hinges on what we mean by "it works" or "it yields satisfactory results" in a scientific context. These results may take the form of resolution of scientific problems, novel predictions, or the stimulation of additional scientific inquiry. Thus, pragmatists are not so much concerned with the truth or falsity of a particular theory (or even of the statements in it), but with the difference that a theory, assumed true, makes to us.

Consider, for instance, the statement "There is a celestial sphere on which all the stars are fixed". Though patently untrue to the non-pragmatist, this statement may be considered true to the pragmatist because of its unquestioned usefulness. We have used and continue to use the apparent positions of the stars for a wide variety of things from night-time navigation to charting out planetary positions. Likewise, a pragmatist may be indifferent to the issue of whether the General Theory of Relativity or Newton's theory of gravity is correct. In many respects, the latter, though replaced by Einstein's theory, has as much right to truth by the principle of Pragmatism in that Newton's gravitational equation and laws of motion are still today used in sending out spacecraft and satellites.

Newton (1642-1727) and the Law of Gravity

In 1687, Sir Isaac Newton published his magnum opus, *Principia Mathematica*. Synthesizing and expanding upon the work of Galileo and Kepler, Newton put forth a dynamical system—his law of gravity and three laws of bodily motion—as binding for all physical objects in the universe. This was the first successful attempt at a universally quantifiable approach to physical phenomena. It was thought to be true for over 200 hundred years, until Einstein put forth the General Theory of Relativity in 1915.

The law of gravity and laws of motion are as follows:

LAW OF GRAVITY ($F = g[Mm/r^2]$): For any two mass points (M & M), the force of attraction is proportional to their masses (M x M) and inversely proportional to the square of the distance between them ($1/r^2$).

1ST LAW OF MOTION: All bodies remain either at rest on in uniform, rectilinear motion, unless impelled by impress forces to change their state.

2ND LAW OF MOTION ($F = ma$): Any change of motion is in proportion to the motive force impressed and remains in the direction of the right line of which that force was impressed.

3RD LAW OF MOTION: To every action, there is always an opposed and equal reaction, or the mutual bodies upon each other are always equal and directed to contrary parts.

The chief problem with the pragmatic theory of truth is obvious: It is not really a theory of truth. To draw out the practical implications of some statement or theory is not the same as to state the conditions under which that statement or hypothesis *would be true*. The latter has predictive and counterfactual force. In addition, according to the pragmatic theory of truth, there seems to be little motivation to take consistency, seemingly a desideratum of any theory of truth, too seriously. Newtonian dynamics may be inconsistent with Einstein's theory of relativity, which today is universally regarded as a better theory, yet Newton's views are still very useful today. By the standard of Pragmatism, if both are equally useful, each can claim an equal stake to being true. This, of course, makes no sense.

Philosopher Michael Lynch sums neatly the inadequacy of Pragmatism as a theory of truth: "What makes a belief useful? If it is just our beliefs, we slide into an unhealthy form of relativism. But if it is the way the world is, we must give up our pragmatist theory of truth".[3]

Key Terms

correspondence theory of truth
Protagorean relativism
Pragmatic theory of truth
coherence theory of truth
cultural relativism (relativistic theory of truth)
Newtonian dynamics

Text Questions

* Of the theories of truth sketched above, which do you consider to be the most tenable or least problematic? What are the chief merits of this view? What are its weaknesses?
* Of the theories sketched above, which is the least tenable and most problematic?

Text-Box Question

What are key differences between Newton's cosmic picture and that of Aristotle, some 2000 years prior to him?

---

1 Aristotle, *Metaphysics*, 1011b25. My translation.
2 Diogenes Laertius, *Lives of Eminent Philosophers,* trans. R. D. Hicks (Harvard University Press, 1991), IX.51.
3 Michael P. Lynch, *True to Life: Why Truth Matters* (Cambridge: The MIT Press, 2004), 67.

Module 6
Realism vs. Antirealism

> "I believe there are only two ways to maintain the antirealist position: either by impoverishing (perhaps I should say emasculating) the methods of science...or by arbitrarily (and vaguely) restricting the scope of application of such principles to the realm of the observable". Clark Glymour, "Explanation and Realism"

SCIENTIFIC REALISM IS SOMETIMES SUMMARILY, AND NAIVELY, SAID to be the view that the practice of science aims at improving our understanding of the way the world is. At first blush, this view seems to be anything but controversial, as it appears to fit squarely with what might be called a common-sense conception of science.

However, as is often the case with common-sense views on issues, once such issues are analyzed with sufficient vigor and thoroughness, what seemed to have been obvious becomes anything but obvious. Within the last few decades, philosophers have begun to challenge this and other naïve views of scientific realism. Antirealists have contested the very claim that science is knowledge-generating. In reply, realists have put forth more sophisticated and less ambiguous accounts of "scientific realism" to defend against these challenges. What makes the realism-antirealism debate even more intriguing is that realists themselves are hopelessly far from a rational consensus on a definition of "scientific realism", which the debate itself presupposes.

Scientific Realism

As scientific realism has many forms—some of which are compatible with relativism and even skepticism—I cannot hope to do justice to all of them in any one definition. As "atheism" is a meaningless term without a clear definition of "theism", "antirealism" also is a meaningless term without a clear definition of "realism". So I begin by offering a general definition of "scientific realism".

> **SCIENTIFIC REALISM is the view (1) that real-world objects exist independently of our ability to perceive them and (2) that scientific investigation goes some way toward disclosing the nature of these objects.**

Let us look at both parts of this definition in turn.

The independent existence of real-world objects is taken to be by most philosophers of science, whether realists and antirealists, as a relatively uncontroversial claim. This commits us to the existence of something external to us that is in some sense causally responsible for our sensory data. When we see trees, penguins, and chairs, we believe that there are certain external objects that are independent of and causally responsible for these perceptions—roughly, real trees, real penguins, and real chairs. We believe that when we pass away, these trees, penguins, and chairs will continue to exist. What is rigorously debated is the nature of these real-world, external objects.

and correspondence theories of truth. For Boyd, scientific terms both map on to real-world objects and are regulated by complex causal interactions with them. Of scientific terms—such as "black hole", "electron", and "aardvark"—he writes, "Roughly, a (type) term *t* refers to some entity *e* just in case complex causal interactions between features of the world and human social practices bring it about that what is said of *t* is, generally speaking and over time, reliably regulated by the real properties of *e*". In short, through investigation, we come to know something about real-world objects, because we act on and are acted on by these real-world objects. What regulates the "real features of the world" is the approximate truth of the background theories or theoretical presuppositions that we have of the world.

Black Holes

According to the General Theory of Relativity, when a body is of a certain mass, nothing, not even light, can escape it. Nearby objects also cannot escape its gravitational pull.

To grasp better what a black hole is, imagine two scenarios. First, imagine throwing a baseball straight up in the air. You throw it, but it always comes back down. Now imagine throwing it so hard that its leaves the earth and is projected into outer space, which is just what happens when we sent spacecrafts to other planets. Next, imagine that the earth is so massive, many times more massive than our sun, that its escape velocity—the velocity an object would need to have to escape the planet—exceeded that of light. Since nothing can travel faster than the speed of light, no object would be able to leave the planet.

That is just what happens on black holes. Imagine that you approach a black hole. As you near its event horizon, you find that you can no longer escape the exceedingly dense object. You move inescapably to its center—called a "singularity". The even horizon then is the place around the black hole where the escape velocity equals the speed of light. Below it, the escape velocity it greater; above it, it is less.

Black holes form from large stars that have burned themselves up. Once burnt up, such stars collapse into themselves and form neutron stars. If the star is very large, the collapse becomes severe and a black hole forms.

The existence of black holes is now well documented by x-ray emissions from x-ray binary stars, where the mass of one star gets sucked into the mass of the other. NGC 4261 and M 87 are examples of objects thought to be black holes.

Let us illustrate by the term "black hole". Black holes are believed to be supernovae explosions where the core matter, left behind after the explosion and generally of the mass of many suns, contracts greatly and is exceedingly dense. Being so dense, the space around the core matter is so warped that light itself can no longer leave. Consequently, the remnants of the core cannot be seen.

Do black holes exist? The existence of such "stars" is predicted by the General Theory of Relativity. That gives us hope that they exist and a good reason to "look" for them. Yet, if they cannot be seen, how are we to know that the term "black hole" is not scientifically meaningless? Though we cannot causally interact directly with black holes, we can and do causally interact with them indirectly. Though black holes are invisible, their existence has certain predictive consequences. Black holes so warp the space around them that any visible body near them will be noticeably perturbed. So black holes are inferable by the behavior of visible massive objects around them (see text box, left).

Overall, for Boyd, there is causal interaction between the real world (i.e., the objects of scientific investigation) and the scientists that interact with it. Scientists use specific terms and verbal descriptions to designate and depict the objects they investigate and to explain their behavior. In such a manner, the causal properties of real-world objects function as mechanisms to correct scientific theories and put them on the path toward a better depiction of the way things really are.[2] For instance, if black holes exist, then continued study and (indirect) causal interaction with such objects will give us more information about them that will enable us to refine existing theories or perhaps develop new theories to better accommodate them. In short, for Boyd, science is a social practice that allows for a meaningful, causal interaction with the real world. In such a manner, science allows us in some measure to know the objects with which we interact.

Policy Realism

A third approach to realism, *policy realism*, is proposed by Rom Harre,[3] who has argued in *Varieties of Realism* that plausible theories—those that predict and retrodict ("predict" backwards) successfully—must have terms that pick out real objects. Science, then, seeks out these objects. Thus, realism is a commitment toward the existence of the entities that scientific theories describe at three distinct levels. At level one, there are *directly observable* entities and those other entities that one can access through simple experiments (e.g., the planets, arteries and veins, and oceanic trenches). At second two, there are entities that are presently unobservable, but *in-principle observable* by humans, given the right equipment. Such entities might, however, require advances in instrumentation to be observed, but the history of science is replete with such advances that have confirmed entities previously thought to be unobservable. For example, the dark side of the moon was once deemed in-principle observable in that, prior to the 1960s, we did not have the technology to send a person to the moon. Likewise, the nearest planetary system to our solar system is in-principle observable, though we possess no direct methods at present of observing it. It would take, for instance, some 70,000 years just to travel to the nearest star, which may not be a solar system. Last, the third level is the existence of entities that are *in-principle unobservable* by humans. Quantum states and Freud's unconscious are good examples. Concerning the latter, Freud repeatedly stressed that the unconscious could be known only indirectly through psychic phenomena such as dreams, slips of the tongue, jokes, and neurosis.[4] Thus, for policy realism, the predictive and retrodictive successes of a theory are themselves sufficient grounds for assuming that the terms picked out by that theory describe real things.

Critique of Scientific Realism

Critics of scientific realism are many. I list merely two of their objections. First, antirealists point to the theory-dependence of scientific statements. Scientific terms and statements have meaning only within the context of a particular theoretical framework that gives them their meaning. Meaning, then, cannot be determined by the fit of terms and statements to the real world independently of that theoretical framework. Therefore, models or theoretical systems of scientific realists may tell us more about individual philosophers or their scientific community than about the nature of reality. A second problem for scientific realists is the notion that the practice of science is a convergence toward truth. If scientists can never be in a position to say that any one theory is true, as no finite amount of evidence can demonstrate that a scientific generalization (seem Hume, Module 7), not limited in scope to past experience, is true, then to what does "convergence to truth" amount?

Scientific Antirealism

Scientific realism, we said, was the view that (1) that real-world objects exist independently of our ability to perceive them and (2) that scientific investigation can go some way toward disclosing the nature of these objects. Thus "antirealism" is simply the view that scientific realism as defined above is false. That, minimally, commits the antirealist to deny one of the two claims of scientific realism. Most deny the second, as it would make little sense to pursue science at all, if one were to deny the first.

There are many different types of antirealism. In what follows, I describe three types: neo-Kantianism, empiricism, and paradigm-relativism. *Neo-Kantians* accept claim one, the independent existence of reality, but maintain that we are conceptually constrained by the nature of our psychic apparatus from getting at the nature of this reality. *Empiricists* also accept claim one, but assert that what we can know or learn about reality comes exclusively through human sensory experience, which is of course quite limited. *Paradigm-relativists,* in contrast, are essentially indifferent to the question of an independently existing reality and focus instead on how scientific practice enriches human living through solving existing problems and anticipating others.

Scientific Neo-Kantianism

Following in the footsteps of the philosopher Immanuel Kant,

> **NEO-KANTIANISM is the view that knowledge of reality is impossible because we interpret what we experience through a conceptual apparatus that moulds or shapes the raw data behind experiences.**

To neo-Kantianism, it is impossible to take in raw, uncontaminated data through experience.

One type of Neo-Kantian view is put forth by the realist-turned-antirealist Hilary Putnam in his more recent work in epistemology and philosophy of science.[5] Once a staunch believer in scientific realism, Putnam has come to adopt a modified form of scientific realism that clings to the notion of independently existing objects, but rejects the notion that we can know anything about them. According to our definition, this makes Putnam's "realism" a form of antirealism.

For Putnam, each person is born with a psychic apparatus that functions by categorizing what it takes in—the raw data of experience. This psychic faculty, what he calls a "commonsense scheme", is roughly the same for each person and allows for certain ways of conceptualizing sensory data and disallows other ways—especially when it comes to certain key conceptions like "object" and "existence".

Putnam gets his argument off the ground by an analogy. He has us consider a cookie cutter and dough. The dough represents the raw data of reality from which perception takes place; the cookie cutter represents the way in which we see reality. This, roughly speaking, is the same for all people. "But", he adds, "this 'cookie cutter' metaphor founders on the question, 'What are the parts of the dough?'" This question does not admit of only one definitive answer.

To illustrate, Putnam has us consider a world with three individuals—x1, x2 and x3. In such a world, it seems obvious that a perfectly logical answer to the question "How many objects are there in this world?" would be "three". Yet consider someone, who thinks that any grouping of two individuals constitutes a distinct object as their sum. To such a person, a perfectly reasonable answer would be seven: x1, x2, x3, x1 + x2, x1 + x3, x2 + x3 and x1 + x2 + x3. Again, he has us consider a plane and its points. Are planes composed of points (as more fundamental parts of a whole) or should points merely be considered to be limits of planes (in which case the planes are fundamental)? The answer to this question, Putnam thinks, depends on whether you conceive of points as entities or limits—how you, as it were, cut the dough.[6]

Scientific Empiricism

Following the tradition of philosophers such as Aristotle, Bacon, Hobbes, Locke, and Hume,

> **SCIENTIFIC EMPIRICISM is the view that all knowledge fundamentally relies on experience.**

The 18th-century philosopher David Hume (1711-1776) has argued that to say "α is the cause of τ" requires that we establish a *necessary connection* between the two events, when all that we are entitled to say from experience is that we have observed the *priority of the cause, their contiguity*, and *their constant conjunction in the past.* No amount of observation, it is obvious, can establish a necessary connection, as observation can only tell us what has happened in the past and a necessary connection is a guarantee that what has happened in the past will happen in the future. Likewise, reason does not solve the problem. No argument can provide a guarantee that the future must be like the past, since it is always conceivable that the future will not be like the past. There is nothing logically impossible, for example, about a universe where the laws of physics change at some point in time.[7] For example, it is entirely possible that, as of the year 2050, the law of entropy (second law of thermodynamics) will reverse itself and things will begin to move toward continually increasing order. For Hume, consequently, the most reasonable attitude concerning the perceived order of visible things is skepticism (see Hume, Module 7). More recently, Baas van Fraasen[8] has put forth a view called *constructive empiricism.* According to van Fraasen, it is the empirical adequacy, not the truth, of a theory that is at issue, and empirical adequacy is a matter of "saving the appearances" as much as possible. For van Fraasen, a scientific concept is "observable" if it can be experienced through the unaided human senses. For instance, the American flag on the moon is observable, since one can travel to the moon to see it. Thus, the statement "There is a flag on the moon" can be shown to be true by a visit to the moon. In contrast, theoretical entities, like sub-atomic particles and Freud's unconscious are not observables, as we can only "see" them indirectly through instruments or methods. In other words, we arrive at "knowledge" of them indirectly through causal inference, like cloud-chamber experiments for submicroscopic particles. Thus, "Submicroscopic particles exist" is certainly a claim that is either true or false, yet, since these particles are not directly observable, no scientist is in a position to assert the truth or falsity of the statement. However, one can, through constructive empiricism, assert that the claim is *empirically adequate* (which it is, since particles leave traces of their trajectories in the "cloudy" chambers), if it accords with observable phenomena, or that it is *empirically inadequate*, if it fails to do so.

"Saving the Appearances" in Greek Astronomy

The notion that a scientific theory is not a commitment toward the existence of the entities posited has had a long history. Following Plato, Aristotle, and perhaps others before them, it was inconceivable that the irregular motions of the planets could be the result of anything other than (1) motion in a circle (2) that was uniform or non-accelerative. Thus, the models early astrographers constructed, in keeping with these two assumptions, were often highly complex.

By the late third century B.C., Apollonius of Perga (fl. 210 B. C.) was perhaps the first to use both epicyclic and eccentric models in an effort to accommodate the planets, whose motions were observed to be anything but uniform over time. It is also likely that Apollonius was the first to demonstrate the mathematical equivalence of these two models.

To illustrate their equivalence, let us assume (Figure 6.1 below) that E represents Earth and P represents a particular planet for both models. Now on the epicyclic model, arm CE moves in uniformly around point E such that point C moves uniformly and circularly around E. Another arm, PC, moves in uniformly around point C such that planet P moves uniformly and circularly around C. From the vantage point of someone on Earth, P's motion seems to speed up as it travels beneath line AB and slow down both AB. Exactly the same apparent acceleration and deceleration of P is explained by the eccentric model. Here one need only assume that the motion is not concentric to Earth, but displaced around point O at some distance from Earth (where PC = OE & CE = PO). According to this model, arm PO moves uniformly around O and planet P then moves in a uniform, circular path.

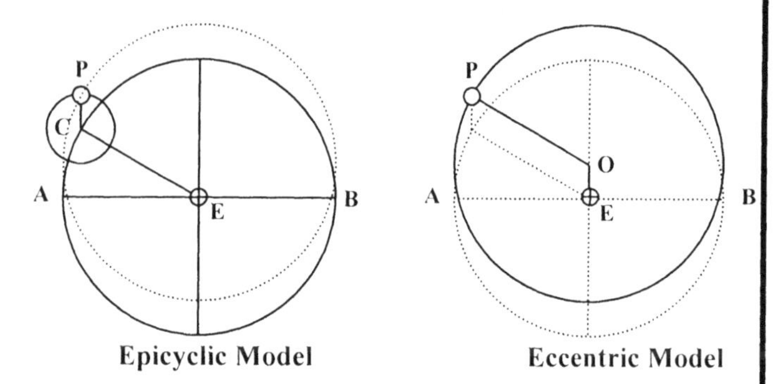

Figure 6.1. Epicyclic and Eccentric Planetary Models. Knowing that there could be two or more mathematically equivalent models that seemingly fit the phenomena made philosophers and astronomers hesitant for centuries to state that such models could do more than "save the appearances".

It is plain to see that canonical empiricism is not a form of realism in any interesting sense. At worst, those who take only sensory data as true or metaphysically fundamental find themselves mired in *solipsism*, a radical form of empiricism where one can be sure only that one's own sensory data are true. So, the solipsist cannot even be sure that there are things, like other people, that are external causes of their sensory data. At best, empiricists admit to the existence of an external reality about which they can have no true understanding whatsoever. In between, there are a host of other problems: their tacit assumption that what is directly observed is not in some measure constructed, their rejection of knowledge independent of experience (i.e., *a priori* knowledge), and the difficulties they often have in relating theory to observation, to name a few. Overall, philosophical explanation today in the empiricist mold is no hunt for causes, but generally a matter of finding an explanation that fits or agrees with observable phenomena. It should, then, come as no surprise that many empiricists are also pragmatists.

Scientific Relativism

Paradigm-relativism, according to Thomas Kuhn,[9] asserts that scientists are not in the business of searching for truths about an independently existing world. For paradigm-relativists, scientific practice is relativistic.

Science, according to the PARADIGM-RELATIVIST APPROACH, is a matter of solving problems that occur within a given scientific community.

Such problems are soluble by employing only those concepts of a community in a manner that is consistent with other acceptable claims within that community. Science, thus, establishes *paradigms* (i.e., patterns or exemplars). Paradigms are consistent ways of seeing the world that both address and solve existing social problems and anticipate other problems in a community. Though consistency within a paradigm is important, it is not essential. The key for paradigm-relativists is that paradigms arise out of other paradigms in answer to problems that prior paradigms could not handle. Though one paradigm arises out of another, the concepts of each paradigm are said to be incommensurable—that is, each has its own language and neither paradigm is reducible to or translatable in terms of the language of the other paradigm. That makes scientific communication between different scientific communities (e.g., earth-centered Ptolemaists and sun-centered Copernicans) strictly impossible.

For paradigm-relativists, the choice between models of reality is not based on rational factors, such as fit with observable reality or explanatory power. Since science is essentially nonrational, it cannot be progressive in any meaningful sense. Science, for them, is essentially a social, not a formal, discipline that answers to the needs of people, not to the dictates of rationality.

One pressing problem of scientific paradigms is the *incommensurability thesis*—Kuhn's notion that distinct paradigms are incommensurable, which means that the language of one paradigm fails completely to match up with the language of another (see Module 12). That however is plainly false. For instance, the language of Copernicanism is certainly not incommensurable with that of Ptolemaism. Each, for instance, relies fundamentally on Aristotelian physical principles, as both theories presuppose the truth of Aristotle's physical principles. Another problem concerns the seeming dismissal of scientific progress. Kuhnians cannot say that the General Theory of Relativity marks an important advance over that of Newtonian Dynamics or even that Newtonian Dynamics is a significant improvement over Aristotle's cosmological principles, only that each was right for its own time in that each came about in answer to the problems of its day. For instance, a paradigm theorist must acknowledge that the technological progress made in medicine is not progressive at all. Such "advances" are merely answering to different and new needs as well as problems in today's medical science that did not exist earlier, which, on sober reflection, seems greatly understated. Moreover, as a paradigm relativist, one could even imagine a society's return to Aristotle's physical principles in preference to those of Einstein and Quantum Theory. That, of course, seems absurd.

KEY TERMS

scientific realism
argument from miracles
policy realism
scientific empiricism
empirical adequacy
scientific relativism
incommensurability thesis
scientific antirealism
causal interactionism
Neo-Kantianism
constructive empiricism
empirical inadequacy
paradigms
Catastrophism

Uniformitarianism
epicylic planetary model
black holes
eccentric planetary model

Text Questions

* Is science in the business of generating truths or merely solving problems? Critically discuss realism and antirealism in your answer.
* In spite of the difficulties with scientific realism, why do some philosophers believe that science enables us to know *something* about the nature of real-world objects?
* In what manner do neo-Kantianism, empiricism, and paradigm relativism each contradict the definition of "scientific realism"?

Text-Box Questions

* What is Catastrophism and how did it come about in response to observation of geological strata?
* Why can we posit that black holes exist, even if they are not in principle observable?
* Why did early astronomers believe that planetary models were only useful approximations and not correct depictions of reality?

---

1 Hilary Putnam, "What Is Realism?", *Scientific Realism,* ed. Jarret Leplin (Berkeley: University of California Press, 1984).
2 Richard Boyd, "The Current Status of Scientific Realism", *Scientific Realism,* ed. Jarret Leplin (Berkeley: University of California Press, 1984).
3 Rom Harre, *Varieties of Realism* (Oxford: Blackwell, 1986).
4 For other "proofs", see his 1915 paper, "The Unconscious".
5 Hilary Putnam, *The Many Faces of Realism* (Chicago: Open Court Publishing Company, 1988).
6 Putnam, *The Many Faces of Realism.*
7 David Hume, *A Treatise of Human Nature*, ed. David Fate Norton and Mary J. Norton (New York: Oxford University Press, 2000).
8 Bas van Fraasen, *The Scientific Image* (Oxford: Oxford University Press, 1980).
9 Thomas Kuhn, *The Structure of Scientific Revolutions* (Chicago: University of Chicago Press, 1996).

SECTION THREE

Causation

Module 7
What Is a Cause?

> "Concerning the first-order question 'Why?', I have raised the second-order question 'Why ask, "Why?"' to which you might naturally respond with the third-order question 'Why ask, "Why ask, 'Why?'"?" But this way lies madness, to say nothing of an infinite regress". Wesley Salmon, "Why Ask, 'Why?'"

AS YOU HAVE PROBABLY FIGURED OUT by now, critical or philosophical analysis of the practice of science is pretty much, in keeping with the quote from Wesley Salmon above, a second-order or meta-discipline. In other words, philosophy of science steps outside of the practice of science to ask philosophically engaging questions such as, "Does science offer a reliable method for helping us make sense of the world?"

The following three modules turn to the issue of scientific explanation through causal analysis of phenomena. This module looks at David Hume's investigation of cause (sketched in Module 6), and then distinguishes between correlative and causal claims. Next, there is a brief philosophical discussion on the issues of determinism and causation. Last, I introduce and distinguish between four types of causal conditions.

Hume on Cause

In his work *Treatise on Human Nature*[1] (1739), the Scottish philosopher David Hume gives an analysis of causal claims.

There are, Hume acknowledges, two accepted forms of reasoning—the first involving relations of ideas (demonstrative or deductive reasoning) and the second involving matters of fact (non-demonstrative or inductive reasoning). He asserts that demonstrative reasoning is a legitimate form of reasoning involving concepts or ideas, whose link with matters of fact is at best tenuous. Non-demonstrative reasoning, in contrast, is generally assumed a legitimate form of reasoning about observable phenomena that gives us conclusions about things unobserved. Yet, Hume asks, what is the foundation of this type of reasoning? To get clearer on this, we must look at the causal relationships that we assert exist between us and the objects we experience.

When we analyze any causal claims, such as "Flame is the cause of heat", we find a commitment to the following conditions:

- the CONTIGUITY of flame and heat,
- the PRIORITY of the flame,
- the CONSTANT CONJUNCTION of flame and heat, and
- a NECESSARY CONNECTION between flame and heat.

Of these, the first three are consistent with experience, but the last condition, in effect, is a requirement that flame *is*, *has always been*, and *will always be* connected with heat. No amount of experience can guarantee this, since no amount of experience can give us complete assurance that the future will be like the past—the very point at issue. So, non-demonstrative or inductive inference is unhelpful. We are left with demonstrative or deductive reasoning—that is, a demonstration that makes no appeal to experience. However, the slightest reflection shows that there can be no demonstrative argument to guarantee a necessary connection among observables, as it is conceivable that the future will not be like the past. This, as we shall see more fully in Module 10, is the problem of induction.

Hume adds that induction is not a legitimate form of reasoning and that it differs from a fiction only insofar as it is conceived differently. Non-demonstrative reasoning, he adds, is a "species of the imagination" and the necessity we ascribe to causal claims that link phenomena is merely a fiction of our imagination. Thus, causal connections are likely not a part of reality, but probably something our minds ascribe to reality. Hume's view of causation is a problem for those who think that science is chiefly or exclusively a bottom-up, inductive enterprise.

Whether Hume's criticism of causation is correct or not, the use of the language of causation is engrained in everyday speaking and it is an important element in scientific explanation.

David Hume (1711-1776)

David Hume was born in 1711 in Berwickshire, Scotland. As a child, he was straightforwardly devoted to Calvinism. At eleven, he matriculated to the University of Edinburgh, which he left some four years later in favor of a private education. His thoughts then turned to philosophy, which became something of an antidote to the Calvinism of his youth.

During these years of private study, Hume was greatly impacted by Pierre Bayle's *Historical and Critical Dictionary* as well as many Greek and Latin authors. He composed and anonymously published his *Treatise of Human Nature* in two volumes (and in two installments, 1739 & 1740). This book, today highly regarded as a philosophical treasure, was received coldly at the time of its publication. Shortly thereafter, Hume published *Essays, Moral and Political* in two volumes and these met with better success.

Around 1744, Hume was being seriously considered for the Chair of Moral Philosophy at the University of Edinburgh. Because of fear of his seemingly anti-religious position on philosophical matters, he was forced to withdraw his candidacy. In 1748, he published *Enquiry Concerning Human Understanding*, which was a reformatting of the content in Book I of his ill-received *Treatise on Human Nature*. Next, Hume published his *Enquiry Concerning the Principles of Morals* in 1751, itself a reformatting and elaboration of Book III of *Treatise on Human Nature*. Here Hume fashioned a morality, devoid of divine precepts, on both utility and human sentiment—a harbinger of Jeremy Bentham's theory of Utilitarianism.

Failure to gain the chair at the University of Glasgow, he secured employment at the Advocate's Library in Edinburgh, where he undertook considerable research in history, which culminated in his widely read *History of England* (published in six volumes from 1754 to 1762). He also wrote *The Natural History of Religion* (1757) and *The Dialogues Concerning Natural Religion* (posthumously, 1779), the latter of which caused considerable religious controversy.

In 1763, Hume became personal secretary of the Earl of Hertford. In 1766, he removed to Edinburgh, where he met with some of the greatest figures of his time—including Jean Jacques Rousseau. Ten years later, at the age of 65, he died. Thought much reviled in his own time, he enjoys a status today as one of the greatest philosophers to have ever lived.

Cause and Correlation

When we examine events that we believe are causally related, say drinking and memory loss, what we actually observe is merely a *correlation* between the variables—that is, that the presence of the one is statistically related to the presence (or absence) of the other (i.e., consumption of alcohol is positively correlated with memory loss). The most obvious explanation for this relationship is causal: One variable (drinking) is a cause of the other (memory loss).

Yet the most obvious explanation in cases of correlation does not always turn out right. Two variables can be correlated without either being the cause of the other. Consider the following scenario. Esther Busburn notices that she is usually depressed whenever it is cold, and so she concludes, "*Coldness* is the cause of my *depression*". As things go, this seems fine, as the two events might be strongly correlated and it would make little sense to say that her depression is the cause of coldness. However, might there not be a third, hidden variable at play here? It just might be that it is not coldness per se that is causing her depression, but *lack of sunlight during winter months*. So it is not only coldness and her depression that are correlated, but also lack of sun and coldness as well as lack of sun and depression. How are we to entangle this mess? The most sensible explanation is that it is the lack of sun that is causally responsible for both the coldness and her depression. The two variables are linked by third variable that is causally responsible for both.

The *Detroit Free Press*, in a radio commercial in 2003, stated that there was a definite positive correlation between *a family's having at least one newspaper in its household* and *the children of the house going on to do well in college*. There is, of course, no reason to doubt that such a correlation exists, as the *Detroit Free Press* is a reputable newspaper. How might one causally explain the correlation? The implicit suggestion here is that the newspaper is the cause of the children's success in college. At least, this is what the advertisers would like you to believe. However, the absurdity of this is obvious. Newspapers alone do not cause children to do well in college. Rather it is that parents, who have great regard for education, tend to have things like newspapers in their household. They also tend to have children who do well in college. Thus, *educated or informed parents*, not newspapers, are causally responsible for their children's success in college (as well as newspapers being in the house).

What the examples above illustrate are essential differences between correlational and causal relationships. To say, for instance, "Flame is correlated with heat" is also to say "Heat is correlated with flame" and this is not to ascribe causal force to either flame or heat. In other words, correlations are reversible relationships between variables. Causal relationships, in contrast, are not. To say "Flame is a cause of heat" is not to say "Heat is a cause of flame". Moreover, to say that two variables are correlated is just to say that two variables tend to be seen together in a manner inexplicable by chance. This indicates that the two variables are related, but it certainly says nothing about just how they are related. Causal relationships, in contrast, seem to be genuine, explanatory relationships between events, where the effect is explicable by the cause.

Are Causes Real?

The view that every event has a sufficient cause is called *determinism*, a working hypothesis of those scientists who assume that all phenomena, at least at the macroscopic level, are regular and law-like. I illustrate with an example from the early history of psychoanalysis.

Two Views of Causality

Historically, there have been two main approaches to causality: causal realism and pragmatic empiricism.

Causal realism has had a long and rich history and can be traced back as far as the early Greek Stoics, of the third century B.C. In a nutshell, it argues that the universe is one big tapestry of items or events that are related to each other, either directly or indirectly, through interactive causal relationships that are in the main extraordinarily complex. In short, the universe as a whole is one colossal deterministic system whose causes as said to "generate" or "bring about" their effects.

The Newtonian physical universe is perhaps still the prime example of such a deterministic system, though Newton himself was rather perplexed about how gravity could exert causal influence from a distance. According to Newtonians, every material body in the universe is gravitationally attracted to every other body—however slight the interaction is. In such a system, any event is in principle predictable with a complete knowledge of the position and momentum of every body at a particular time as well as the laws of physics. A change in the position or momentum of any one body at any time has an impact, even if negligible, on the entire system. Wrote Pierre Simon de Laplace over two hundred years ago of these implications:

> Present events are connected with preceding ones by a tie based upon the evident principle that a thing cannot occur without a cause which produces it. This axiom, known by the name of the principle of sufficient reason, extends even to actions which are considered indifferent; the freest will is unable without a determinate motive force to give them birth....
>
> We ought then to regard the present state of the universe as the effect of its anterior state and as the cause of the one which is to follow. Given one instant an intelligence which could comprehend all the forces by which nature is animated and the respective situation of the beings who compose it—an intelligence sufficiently vast to submit these data to analysis—it would embrace in the same formula the movements of the greatest bodies of the universe and those of the lightest atom; for it, nothing would be uncertain and the future, as the past, would be present to its eyes.

Causal pragmatic empiricism, which is perhaps the predominant view among philosophers today, was profoundly impacted by the skepticism of Hume. Taking experience as his guide, Hume argued that philosophers and scientists who talked about causal relationships were never in a position to assert more than the observed contiguity and constant conjunction of "cause" and "effect" as well as the observed priority of the putative cause. What could not be guaranteed by observation was a necessary connection between the two.

Pragmatic empiricists today have been inspired not only by the skepticism of Hume, but also the challenges to causal realism that have come from modern science—especially Quantum Physics (see Modules 18 and 20, text boxes). Using the evidence of our senses, pragmatic empiricists assert that we are never in a position to posit a necessary connection between what we call "causes" and "effects". Ascription of causal force is then a projection on to reality that can never be fully justified. Causal language is thus no more than a useful fiction that guides investigative empirical inquiry.

In 1880, the Viennese doctor Josef Breuer took on a hysterical patient as favor to her family—a 21-year-old, precocious female who was called "Anna O.". In July of 1880, Anna's father became very ill and bedridden. Anna was distraught and became his primary caretaker. Her distress over this situation took its toll shortly. Having symptoms that included severe coughing, squinting, hindered vision, headaches, paralysis of her neck and arms, she became physically weak. At times, she could talk and write only in English, as if ignorant of her native German. Her condition deteriorated even more upon the death of her father in April of 1881.

Breuer, in his sessions with Anna, discovered that if she, under hypnosis, talked about her hallucinations from a particular day, she would enter a state of tranquility. When there was no such recitation during a session, she would be extremely anxious. Anna herself took notice and began to refer to these therapeutic chat sessions as "chimney sweepings".

At one point, Anna suddenly became hydrophobic. Her hydrophobia persisted for six weeks, in which she ate fruits for fluid. During one of her chimney-sweeping sessions with Breuer, she mentioned an Englishwoman, whom she hated. Anna spoke of this woman once having let her dog drink water from a glass. Anna related this event with intense feelings of revulsion. Once the episode was related to Breuer, Anna asked for a glass of water form Breuer. Immediately, her symptoms disappeared, never to return.

In his early collaborative work with Breuer, Sigmund Freud learned that even psychical phenomena, like Anna's hydrophobia, were effects of prior causes. In the case of Anna, her symptoms, Freud came to believe, could be traced back causally to unpleasant and forgotten events in her life. Freud's work with Breuer was setting the foundation for his causal theory of repression, the key conception of psychoanalysis, in later years.

The example above illustrates that scientists' work is based on the assumption of *determinism*—that all observable events, at least at the macroscopic level, have sufficient causal conditions that bring them about. For Freud, humans themselves, even at the psychical level, are deterministic systems, causally linked in complex ways to the things around them.

Are causes real things that operate in the world of observable things? Many philosophers think so. Others, pragmatically inclined, take causation merely as working hypothesis—a useful assumption that may or may not be true of world. Either way, the notion of "cause" is an indispensable (or very nearly so) part of scientific explanation.

For the purpose of this undertaking, let us assume a position of neutrality on the metaphysical status of causes. Though I shall often speak of phenomena in terms of causal relationships that actually exist in reality hereafter, the commitment is not metaphysical, but pragmatic.

Four Types of Causal Conditions

Below, I list four types of causal condition: a sufficient causal condition, a necessary causal condition, a sufficient and necessary causal condition, and a probabilistic causal condition.

Sufficient Causal Condition

For any two events, α and β, α is a SUFFICIENT CAUSAL CONDITION for β, if, whenever α occurs, β must occur.

Example:

Beheading a human being is sufficient for death.

A sufficient causal condition asserts that every time the causal condition is present, so too will the effect be present. The example above asserts that no human can exist with his head lopped off and this, of course, seems reasonable.

Necessary Causal Condition

For any two events, α and β, α is a NECESSARY CAUSAL CONDITION for β, if, when α does

not occur, β cannot occur.

Example:

Oxygen is necessary for flame.

A necessary causal condition is a condition that must be present for a certain effect to occur. There can be no flame without oxygen.

Sufficient and Necessary Causal Condition

For any two events, α and β, α is a SUFFICIENT AND NECESSARY CAUSAL CONDITION for β, if, when α occurs, β must occur and when α does not occur, β cannot occur.

Example:

Experience of accelerated motion is a sufficient and necessary condition of experience of a gravitational field.

Here the experience of accelerated motion is sufficient and necessary for experience of a gravitational field, since, according to the General Theory of Relativity, there is no testable difference between the two. Einstein himself gives the example of an observer, who experiences a jerk forward while riding on a railway carriage. This, he says, could be the result of the application of the carriage brake, but it could also be fully explained by the existence of a forward-directed gravitational field.[2] According to relativity, acceleration is equivalent to gravitational attraction.

Probabilistic Causal Condition

For any two events, α and β, α is a PROBABILISTIC CAUSAL CONDITION of β, if, when α occurs, it increases or decreases the likelihood of β.

Example:

Heavy drinking is a probabilistic causal condition of memory loss.

It is important to note here that ascription of probabilistic causal force, at least at the level of observable phenomena, like the rolling of dice, is a measure of one's ignorance of determining factors (i.e., the movement of the dice in the shaker's hand, the weight of the dice, the force with which they are thrown, the surface of the table upon which they are thrown, etc.). In such a scenario, the best one can do is to calculate—using past experience or a priori considerations as one's guide—the likelihood of a desired outcome. This, of course, can never guarantee that this desired outcome will come about. Probability, however, does seems have genuine causal force at the level of subatomic phenomena.

KEY TERMS

Hume on cause
determinism
causal pragmatic empiricism
necessary causal condition
probabilistic causal condition
correlation
causal realism
sufficient causal condition
sufficient and necessary causal condition

TEXT QUESTIONS

* What are the differences between a correlation and a causal relationship between variables?
* Why does David Hume think that the necessity behind causal claims is a fiction?
* What are the four different types of causal conditions?
* Why are the probabilistic causal condition important even though phenomena at the macroscopic level are thought to be deterministic systems or mostly so?

TEXT-BOX QUESTION

What is the difference between causal realism and causal pragmatic empiricism?

EXERCISE

Using the example below as a guide, identify the following as instances of sufficient causal condition, necessary causal condition, sufficient and necessary causal condition, or probabilistic causal condition. In ambiguous instances, choose the best answer.

E.g.: *Brushing one's teeth* and *having fewer cavities.*

Brushing one's teeth is a probabilistic causal condition of *having fewer cavities.*

1. *Reckless lifestyle* and *premature death.*

____.

2. *Studying science* and *becoming a scientist.*

____.

3. *Using a telescope* and *seeing the rings of Saturn from Earth.*

____.

4. *Being married* and *celebrating one's 25th wedding anniversary.*

____.

5. *Putting molecules in motion* and *generating heat.*

____.

6. *Removing someone's brain* and *his loss of consciousness.*

____.

7. *Regular exercise* and *physical fitness.*

____.

8. *Drug use* and *drug addiction.*

____.

9. *Stroke* and *paralysis.*

____.

10. *Warping space* and *creating a gravitational field.*

____.

1 David Hume, *A Treatise of Human Nature,* ed. Ernest C. Mossner (Viking Press, 1986).
2 Albert Einstein, *Relativity: The Special and General* Theory (New York: Bonanza Books, 1961).

Module 8
Mill's Methods

"If one tries to describe processes of genuine thinking in terms of formal traditional logic, the result is often unsatisfactory: one has, then, a series of correct operations, but the sense of the process and what was vital, forceful, creative in it seems somehow to have evaporated in the formulations". Max Wertheimer, *Productive Thinking*

BEGINNING IN THE 12TH CENTURY, many of Aristotle's works on science and its method became accessible in Latin to scholars in Europe for the first time. This brought about a revolution in scholarship by capable scholars such as Robert Grosseteste (c. 1168-1253), Roger Bacon (c. 1214-1292), John Duns Scotus (c. 1265-1308), and William of Ockham (c. 1280-1349), each of whom contributed significantly to Aristotle's account of the inductive phase of the logic of science. Later scholars—such as Francis Bacon (1561-1626), John Hershel (1792-1871), and William Whewell (1794-1866)—would expand significantly on these insights.

Building upon this work, in *System of Logic* (1843), John Stuart Mill (1806-1873) proposed what have come to be known Mill's Methods—five inductive methods that he believed could discover real causal connections among visible phenomena. Mill, at times, made outlandish claims for these methods. In *System of Logic*, he claims that every causal law that has hitherto been discovered has been discovered by use of one of these methods. Though he acknowledged difficulties in their application, as when a particular effect is due to more than one cause, he was in the main a staunch believer in their overall effectiveness.[1]

Method of Agreement

By looking at a particular event that stands in need of causal explanation, the *method of agreement* groups together all cases of the event in question as well as all potentially relevant circumstances prior to the event, and then looks to see if there is any one element of these circumstances that is common in each case or the overwhelming majority of them. If so, there is some reason to believe that this element is the cause of the event under investigation. Consider the following example as a means of illustrating.

On a sunny day in July, 515 passengers take off on a cruise ship to Kamchatka. By day four, 151 of the passengers become ill. Symptoms include stomach cramps, diarrhea, and vomiting. Of those who are ill, the ship's director, Angie Prawn, collects what data she can in an effort to discover the cause of the illness.

Cases	*Prior Circumstances*	*Event under Investigation*
Passenger 1	S, O	E = Illness
Passenger 2	E, O, T	E = Illness
Passenger 3	S, T, O, E	E = Illness
.	. .	
Passenger 151	T, O, E	E = Illness

John Stuart Mill (1806-1873)

John Stuart Mill was born in London on in 1806 and was educated by his father, James Mill, who exposed him from early life to a rigorous system of study. His childhood was then anything but normal. Mill himself recognized that his early intellectual training left no time for him to explore a healthy emotional life, though it did give him a decided advantage over other scholars of his day.

Mill learned Greek, arithmetic, geometry, and algebra by the time he was eight. He began reading Plato at seven and, at such time, also began tutoring his sister. At eight, he began Latin and turned to Plato's more important dialogues. He then turned to Homer, Virgil, Horace, Livy, Sallust, and Ovid. At 11, he helped his father in correcting proofs of the latter's *History of India.* At 12 he began to study logic. At 13, he studied political economy and he went to France at 14 to further his education.

With no time for anything but study, he was friendless as a boy. At 17, his father appointed him his subordinate at the East India Company, where Mill remained until the company folded in 1858. This at least allowed him sufficient leisure to pursue his wide range of interests.

Due to continual study, a sheltered childhood, and an authoritarian father, he suffered a mental breakdown at 21. His life he judged meaningless. He deemed himself to be a thinking machine, without feeling and passion. He found solace in poetry, which gave him some outlet for pent up emotions.

In 1830, he met Harriet Taylor, with whom he fell in love, although she was married. Her husband John died in 1849 and Mill married Harriet Taylor two years later. When Harriet died in 1858, he bought a cottage overlooking the cemetery in which she was buried.

Mill published *System of Logic, Ratiocinative and Inductive* in 1843. In the following year, he published *Essays on some Unsettled Questions of Political Economy* and, in 1848, *Principles of Political Economy.* In 1859, he published his groundbreaking treatise *On Liberty* as well as *Thoughts on Parliamentary Reform. Utilitarianism* came out in 1863. In 1869, his treatise *The Subjection of Women* was published. He died at Avignon in 1873. Thereafter, his *Autobiography* and *Three Essays on Religion: Nature, the Utility of Religion, and Theism* were published.

After grouping together all reasonable causes under the prior circumstances (S = use of unsanitary bathroom on deck two, E = consumption of egg-based bread pudding on day one, O = consumption of oysters at the "Oyster Festival" on day two, T = consumption of tuna on day two), Angie looks to see whether any one suspected cause is common to all (or nearly all) cases of illness. Let us say that she discovers that 149 of those ill state that they attended the Oyster Festival and claimed to have eaten oysters on day two of the trip. By the method of agreement, she is in position to claim that *consumption of oysters caused the illness.* Angie tests the remainder of the oysters and finds that most have some measure of fecal contamination. She throws them overboard.

As seems sensible, this method is most effective when there are a large number of cases that are inspected. Of course, this makes the data-gathering part most difficult. Another problem with the method is that finding something common to all scenarios does not necessarily show this to be the cause or the *only* cause of the event under investigation. There is always the possibility of some causal factor that was overlooked. This method, at best, serves as a springboard for further investigation.

Method of Difference

The *method of difference* is strictly a comparison of two cases—one in which the effect of some unknown cause is present, the other in which the effect is absent.

Let us illustrate by imagining a celebrated British female sprinter, Cynthia Celer. Cynthia, having won the silver medal at the Summer Olympics of 2000, had performed exceptionally well, but performs poorly at the Olympics in 2004. Not only does she not place, but her time is significantly off that of 2000. Having had full expectation of performing better in 2004, as she was quite young in 2000, she has good reason to ask: Why was there such a big difference in 2004? She jots down a list of all relevant circumstances (H = healthy diet, T = solid training sessions, I = injury-free training; & D = use of performance-enhancing drugs).

Cases	*Prior Circumstances*	*Event under Investigation*
2000	H, T, I, D	E = Success
2004	H, T, I, ~D	~E = Failure

Cynthia notes that her diet, training, and physical health were just as good as, if not better than, they were in 2000 just prior to the Olympics. She comes to realize that the only relevant difference between the four years is her failure to use the performance-enhancing drugs that she had experimented with four years ago, just one month prior to the Olympics. She concludes that her *use of performance drugs is the cause of her better performance in the prior Olympics.*

Like the method of agreement, this method too has its limitations—not the least of which is that the real cause must be among the list of circumstances jotted down for this method to work. This, of course, requires an exhaustive investigation of potential causes at the outset, which is always difficult to do.

Joint Method of Agreement and Difference

A third method, which functions as a check for the method of agreement, is the *joint method of agreement and difference*. Here there is a comparison of many cases in which the event under investigation occurs with many cases in which the event under investigation does not occur. This is roughly the same method as the method of difference, with the exception that two groups of cases are contrasted, not just two cases. We look for one causal factor that is present in each (or the overwhelming majority) of the events where the effect occurs and absent from each (or the overwhelming majority) of the events where the effect does not occur.

I return to the mysterious illnesses on the cruise ship to Kamchatka to illustrate. Here I list again all cases where illness occurred.

Cases	*Prior Circumstances*	*Event under Investigation*
Passenger 1	S, O	E = Illness
Passenger 2	E, O, T	E = Illness
Passenger 3	S, T, O, E	E = Illness
	. . .	
Passenger 151	T, O, E	E = Illness

So far, this is no different from the method of agreement. Now, however, let us contrast this with all cases where illness did not occur, which is in effect the use of the method of difference to two groups of cases, instead of two individual cases.

linked to particular events under investigation, say w, x, & y, and if one can show W is the cause of w and X is the cause of x, it then follows, by elimination, that Y must be the cause of y.

To illustrate, J. Abraham Zmidlewski discovers that his blood pressure is abnormally high: While usually it is 120/80, it is now 170/110—an increase of +50/+30. Thinking about all the possible causes of this change in blood pressure, he realizes that he has made two significant changes to his lifestyle in the past year and he wishes to find out how each of these is causally related to his heightened blood pressure. First, after years of being unemployed, he has gotten a job. Second, as a result of his new job, he has begun smoking. Which, he asks, is responsible for his higher blood pressure? Are both?

As a test, he cuts out smoking altogether and notices a sharp decline in his blood pressure, though it is not back to normal. It is now 150/100, so he has accounted for +20/+10. He concludes that his new job is responsible for the rest of the high blood pressure—the +30/+20 he has not accounted for.

Key Terms

method of agreement	method of difference
joint method of agreement & difference	method of concomitant variation
method of residues	Skinner on Ψ-states

Text Question

What are the strengths and weakness of each of the five causal methods of Mill?

Text-Box Question

Why does Skinner believe that inner states play no role in psychological explanation?

Exercise

Identify each of the following as one of the five causal methods of Mill: method of agreement, method of difference, method of agreement and difference, method of concomitant variation, or method of residues. In difficult cases, work backwards from effect to possible causes.

1. Alexander Hood has been going out with "boys" a lot more these days than he has in the past. He also notices that he's been sleeping less soundly of late. What is particularly striking, he notes, is that he seems to sleep the least soundly on those nights that he drinks the most. What method is he using?

Cases (Groups) *Prior circumstances* *Event under Investigation*

2. Garruth V. Narcissus is distraught, as the fourth straight woman he's been dating has just broken up with him. He recalls some of the criticisms of him each woman has had over the years. His first, Pamela, claimed he was pigheaded, self-centered, and slobbish. Ursula, his second, stated he was undereducated, fat, self-centered, and unemployable. Sandy, his third, said he was uninteresting, chubby, boring, and self-centered. Finally, Sabrina said he was tight with cash, self-centered, and unhappy. Garruth concludes that, if he is to keep a woman, he must do something about his self-centeredness. What method did he use?

Cases (Groups) *Prior Circumstances* *Event under Investigation*

3. Aged Mr. Rodney E. Putin pledges to begin taking the new multi-vitamin that his wife has purchased for him. The last multi-vitamin he took listed merely an antioxidant capacity. The new multi-vitamin lists an antioxidant capacity as well as a capacity for improved sex drive. The new multi-vitamin lists all the same vitamins and minerals that the old one listed with the exception of ginseng. Aged Mr. Putin concludes that the ginseng must be responsible for the capacity to improve sex drive.

Cases (Groups) *Prior Circumstances* *Event under Investigation*

4. When she last traveled through Chicago's O'Hare Airport years ago, Annette Ayer-Gant was not stopped and searched, but this day she is. As she was then neatly dressed, of fine bearing, and of cheery countenance, as she is now, she concludes that her being searched must be because of security concerns at major airports due to heightened terrorist alerts that did not exist years ago.

Cases (Groups) *Prior Circumstances* *Event under Investigation*

5. While going through her grades at the end of the term, Professor Kathryn "Kay" Matheson noticed that, of all her students, those that fared well all had exceptional attendance and those that fared badly all had poor attendance. She concluded that attendance and its lack were the causes of success and failure in her class.

Cases (Groups) *Prior Circumstances* *Event under Investigation*

---

1 J. S. Mill, *System of Logic* (London: Longmans, Green, 1865).

Module 9
Causal Fallacies

"Oh, the lives that are spent in senseless blunderings,
The ingenuity, the invention, the eternal schemes—
All wasted for the want of one small scrap
Of butterfly knowledge—how to teach fools
To be wise, and think". Euripides, *Hippolytus*

SINCE CAUSAL REASONING PLAYS SUCH A LARGE ROLE in science, it is necessary to be able to recognize when causal reasoning goes awry. This module lists some of the most important causal fallacies. Knowing these fallacies is a critical part of evaluating arguments with causal claims.

Post Hoc (ergo Propter Hoc) Fallacy

Literally, this is the fallacy of "after this (so, because of this)". This fallacy occurs when one attributes causal force to something simply because it occurred prior to something else that is seen to be connected with it.

Example:

> Kalamazoo pitcher Rusty Rooten lost his glove before a game and had to borrow a glove from a teammate before pitching. Rusty then pitched a no-hitter. After the game, he claimed that the borrowed glove was the reason he pitched a no-hitter.

The fact that one thing (here, putting on a different glove) occurs before some other thing (the no-hitter) is itself insufficient to demonstrate an actual causal relationship. Temporal succession of events alone does not establish a causal relationship between them.

Ignoring a Common Cause

This fallacy occurs when one falsely attributes an effect of a cause as a cause for another of the effects of the same cause.

Example:

> Families who have newspapers delivered to their house have children who do better in college. Therefore, newspapers in a house are causally responsible for the children doing better in college.

Here, the parents are causally responsible for newspapers delivered to their house and their children doing well in college.

Confusing Cause and Effect

In a causal relationship of two events, the fallacy of confusing cause and effect comes when one judges the cause to be the effect and, conversely, the effect to be the cause.

Example:

> People who supplement their diet with synthetic vitamins tend to be healthier than people who do not. Therefore, vitamins are a cause of improved health.

It may be that this is a legitimate and strong argument. It may also be the case that this is an instance of confusing cause and effect, as it is quite possible that synthetic vitamin supplementation does not improve health. It may just be that healthier people tend to take vitamins more often than those who are not so healthy. For instance, some research on the use of vitamin E suggests that it may be harmful, if taken synthetically. Further research on the impact of synthetic vitamins on health needs to be done.

Logic and the Lottery

Is there a scientific approach to playing the lottery? The easy answer is "no". Let us look at a simple scenario.

Consider a three-digit lottery, where for the price of one dollar, Amyr gets a chance at $500 by picking all three digits. Amyr, by picking 5-6-6 (or any combination, for this matter), has a 1/1000 chance of each digit being correct (i.e., 1/10 x 1/10 x 1/10), as each of these digits coming up is independent of the others coming up. Now suppose, over the course of the first two months of the year, Amyr notices that the number "2" has come up eight times more than the next most frequent digit and the number "9" has come up the least often in the same period of time. Now Amyr seems to have one of two reasonable strategies at his disposal. Using past experience as a guide, he might play only number sequences that have at least one number "9" in them. On the other hand, noting that "2" has come up less often than other numbers, he might adopt a compensatory strategy and choose to play sequences with the number "2". To adopt either strategy is to commit the gambler's fallacy.

Suppose further that Amyr has tried and failed at both strategies. Frustrated, he wants a foolproof strategy that guarantees he will win the three-digit. It occurs to him that there are ten possibilities for each number, which lead to 1000 possible outcomes (10 x 10 x 10) prior to the numbers being chosen by the lottery committee (i.e., [0 & 0 & 0] or [0 & 0 & 1] or...or [9 & 9 & 8] or [9 & 9 & 9]). He reasons that all he has to do is to purchase one ticket for each of the 1000 possible outcomes to win the $500. He congratulates himself on having solved the mystery of the three-digit lottery, purchases a ticket for each sequence, and wins $500 on the day of the drawing. Success is short-lived, as he comes to realize that he has just spent $1000 to win $500.

Gambler's Fallacy

This happens when two causally independent events are assumed to be causally dependent.

Example:

> Poker player Al "The River" Stackhouse has been getting great cards all night long. Consequently, his luck is bound to run out soon.

The cards that Stackhouse will get in the future are not affected by the cards he has been getting thus far. Thus, it is ridiculous to conclude that his luck will run out because of his lucky run.

The Story of Charles "Victory" Faust

Charles Victor Faust or the "he Kansas Jinx Killer" was born in Marion, Kansas. At the age of 30, he left his family and farm to go to Wichita. Here he visited a fortune-teller, who said to him that he would join the New York Giants, become the greatest pitcher of all time, and help the team to win the pennant.

On the morning of July 28, 1911, while the Giants were in St. Louis to play the Cardinals, Faust approached their manager, John McGraw, and told McGraw of his incident with the fortune-teller. McGraw allowed Faust to practice with the team. Faust was dreadful, but McGraw allowed him to sit on the bench for the game later on, and the Cardinals won 5-2. The next day, Faust, fitted sloppily in a Giants uniform, came back with the team and the Giants won 8-0. Impressed, McGraw let Faust travel with the team. Before each game, Faust would warm-up as if he would be the starting pitcher and then take his spot on the bench. New York went on a winning streak and McGraw's middle name was changed to "Victory".

The Giants did manage to win the pennant that year by seven and one-half games, but they lost the World Series to the Philadelphia Athletics, four games to two. McGraw would later write of the season, "I give Charlie full credit for winning the pennant for me—the National League pennant of 1911".

Faust was back the following season to sit on the bench, though he was not given a uniform. Again the Giants won the pennant and failed to win the World Series. Precisely the same thing occurred the following year. McGraw grew tired of this scenario and did not invite Faust back for the 1914 season, where the Giants finished in second place. Faust was committed to a mental institution in Fort Steilacoom, Washington where he died of tuberculosis on June 18, 1915. That year the Giants finished in last place!

Likewise, there can be no guarantee that it will continue. Probability, as it is sometimes said, does not work by compensation, but rather by swamping.

Genetic Fallacy

This fallacy occurs whenever someone argues that a claim is true or false because of its origin.

Example:

> Freud's idea of the Oedipus complex came from a myth, so there is nothing to the Oedipus complex.

As we mention earlier, that a scientific idea comes to someone in a strange way is not sufficient proof that the idea is wrong. Scientists draw inspiration from the strangest places.

KEY TERMS

post-hoc fallacy
confusing cause and effect
genetic fallacy
ignoring a common cause
gambler's fallacy

TEXT QUESTIONS

* The post-hoc fallacy is extraordinarily common in professional sports. Can you identify one or more instances of this fallacy in sports with which you are familiar?

* The gambler's fallacy is also extremely common in everyday reasoning about sports statistics (e.g., arguing in soccer that a particular player will score a goal because he has been on a scoring draught of late). How often to you find yourself committing this fallacy, when watching a particular sport?

TEXT-BOX QUESTION

What is a sure-fire way to win the three-digit lottery? Why is winning in that manner still a bad bargain?

EXERCISE

Identify the appropriate causal fallacy in each of the following arguments. In difficult cases, choose the best answer.

1. After the funding for his research had been cut, Dr. Lud T. Pittleton became angry and quit his job at Bugitcutt University. His former colleague reasoned that his anger caused him to quit his job. What fallacy did his former colleague commit?

____

2. Many of the exaggerated views that people have of their own fathers are generated from the view that god is an exalted father figure. Since these exaggerated views do more harm than good, it follows that god cannot be an exalted father figure.

____

3. The last 10 experiments performed in Strong Laboratory have failed to yield any worthwhile results. Therefore, the very next experiment is almost certain to be pointless.

____

4. Benjamin Cox carried his father's old fountain pen in his coat pocket on the day of his interview for the job at Ajax Truck Lines. When he got the job, he argued that the pen made the difference. What fallacy did Benjamin commit?

____

5. There is more information about sex-related diseases than there are instances of sex-related diseases. Therefore, the incidence of sex-related diseases is caused by widespread information concerning these diseases.

____

SECTION FOUR

Science & Progress

Module 10
Experience & the Possibility of Knowledge

"As far as the laws of mathematics refer to reality, they are not certain; and as far as they are certain, they do not refer to reality". Albert Einstein, "Geometry and Experience".

Scientific A-Priorism

Perhaps in response to the revival of Aristotelianism beginning in the twelfth century and early inductivist approaches to science, there began attempts to resuscitate deductivist approaches to science beginning in the late sixteen century. In the sixteenth century, Pythagoreans and Platonists saw nature as a book, written in mathematical language. Followers of Archimedes and Euclid viewed science as a branch of metaphysics, where the principles of science were, in effect, thought to be deducible from the most intuitively obvious first principles of reality.

Kepler's stubborn belief that God, in the construction of the universe, was a Pythagorean enabled him through sheer persistence to discover the three laws of planetary motion (see Module 4), centering on the idea that the planets moved elliptically, not circularly. In *Mysteries of the Cosmos*, he tried to show that the planetary distances varied in proportion to the radii of each of their planetary shells, each of which was thought to be inscribed within one of the five "regular" solid, nested figures—the cube, the tetrahedron (four-sided solid), the dodecahedron (12-sided solid), the icosohedron (20-sided solid), and the octahedron (eight-sided solid) (see Figure 10.1).

Figure 10.1 Kepler's Planetary Shells

Like Kepler, Galileo Galilei (1564-1642) worked with a firm belief that a true grasp of the way things worked could only be one that was mathematical. He laid the ground for Newton's law of inertia (see text box, p. 92) in stating that a body moving on a frictionless spherical plane, like a ball rolling across a frictionless globe, would have no reason to stop.[1] The frictionless plane, he noted, was fictive, as a body moving on a plane must create some friction.

Some of his most important work concerned falling bodies. He noticed that, while rolling balls down inclined planes of equal height but varying inclinations, the speeds of the balls at the bottom were the same in each case. He posited that in a vacuum, another frictionless experimental condition that could not be duplicated in the real world, bodies moving downward (that is, toward the center of the earth) would travel a distance in proportion to the time squared ($d \propto t^2$). Thus, he discovered the law of falling bodies.

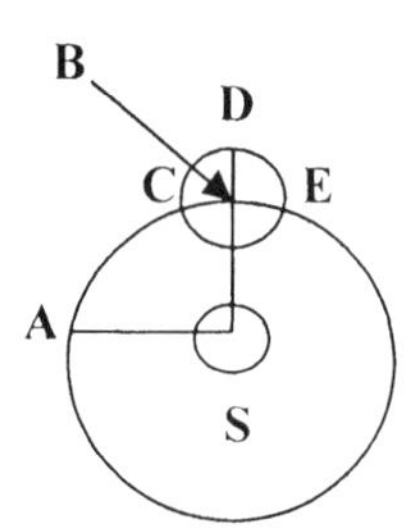

Figure 10.2 Galileo's Theory of Tides. Suppose S is the center of the earth's orbit, the circle around S is represents the path of the earth's yearly motion around the sun (west to east, in the direction of B to A, counterclockwise), and the circle around B is the earth itself (here greatly exaggerated), which moves on its own axis (west to east, in the direction of E to C, counterclockwise). For someone experiencing night (someone moving from E to C) the effect will be additive, as the earth will be spinning in the same direction as the motion of earth in space; for someone experiencing day (someone moving from C to E), the effect will be subtractive, as the earth will be spinning in a direction contrary to the motion of the earth in space.

Such claims show plainly that Galileo's theoretical intuitions, appealing to ideal conditions like frictionless surfaces, went beyond observation and could not be perfectly replicated or fully confirmed in physical reality. At times, when observations failed to agree with his theoretical claims, he faulted the observers and their measuring instruments rather than his own theoretical intuitions. It comes as no surprise that Galileo often stubbornly held on to hypotheses, like his theory of tides (thought to be a sort of back-and-forth swishing of water caused by both the earth's yearly and constant motion around the sun and the earth's daily and constant rotation on its own axis; see Figure 10.2), shown by observation to be wrong, by positing other "secondary" causal factors, like coastline irregularities and variations in the depth of the sea.

Also committed to a mathematical approach to natural phenomena, evident in his universal law of gravity and three laws of bodily motion, was Isaac Newton. Though he was committed to an approach to science that combined both inductive and deductive reasoning, especially in his work on light and color in *Opticks*, it is clear that Newton's laws of motion (see Module 5) could not have been "inferred by general induction from phenomena".[2] The law of inertia, for instance, states that a body, unperturbed in its motion or rest in empty space, will remain in motion or at rest. Now given that gravity requires every body everywhere in the universe has some impact, however remote, on every other body, what *observations* could Newton have made to infer and render general the law of inertia? None. In addition, law two states that a particular change of motion from an inertial path is in proportion to the impressed force responsible for the change of motion. Again, how is one to measure precisely such an impressed force given the interference of impressed forces of all other bodies in the universe on this body changing its motion? The same general criticism holds for law three. Newton's theoretical scaffolding was not the result of inductive inference based on meticulous observation, but rather was built upon the work of predecessors by sharpening and correcting their views and putting them into an axiomatic system, in the manner of Euclid and Archimedes, with heretofore unthinkable explanatory power. His laws were said to reign over all bodies everywhere in the universe. He never denied, of course, an inescapable empirical dimension—that the ultimate test for all theoretical claims was their fit with experience, but his system was, in key respects, non-empirically devised.

Galileo and the Law of Inertia

Using Aristotle's physics, Copernicus' heliocentric system was not less complex than that of Ptolemy. For one, if the earth moves around the sun and spins on its axis (at around 1000 m/h at the equator), why is it that bodies that leave the earth, like birds or projectiles, are not rapidly displaced?

Galileo had to assume something similar to Newton's principle of inertia (in empty space, a body in straight-line motion will remain in motion and a body at rest will remain at rest) to get Copernicanism off the ground. In *Dialogue Concerning Two Chief World Systems*, Salviati, the Copernican, has Simplicio, the Aristotelian, consider a ball, on the surface of a frictionless plane, given an impetus in some direction:

> Sim.: I cannot see any cause for acceleration or deceleration, there being no slope upward or downward.
> Sal.: Exactly so. But if there is no cause for the ball's retardation, there ought to be still less for its coming to rest; so how far would you have the ball continue to move?
> Sim: As far as the extension of the surface continued without rising or falling.
> Sal: Then if such a space were unbounded, the motion on it would likewise be boundless? That is, perpetual?
> Sim.: It seems so to me, if the movable body were of durable material.
> Sal.: That is of course assumed, since we said that all external and accidental impediments were to be removed....

So moving bodies on a spherical surface, like the earth, without friction, would continue to move circularly without end. This motion is independent of another tendency to move, in a continually accelerated manner, toward the center of the earth.

To show the independence of the two motions, Galileo gives another thought experiment. Salviati has Simplicio consider a ship (Figure 10.3), sitting at rest in the sea, and a stone dropped from the top of its mast. The stone, falling parallel to the mast (from Q), quickly drops to the foot of the mast (to P). Now consider the same ship traveling smoothly in the sea in some direction. When the stone is dropped (from Q), Simplicio asserts, it will fall some distance behind the mast (at R) because the ship will have moved forward. That, he asserts, has been observed on numerous occasions by certain authorities. Salviati states that no such observations could have ever been made, since the stone on the moving ship must fall to the foot of the mast (to P, Figure 10.4), as if the ship were motionless. More than this, it will actually fall slightly in front of the foot of the mast (to F, not P), as the circular motion of one who drops the stone at the top of the mast will be slightly greater than the circular motion of the ship at the bottom. In other words, one who participates in the ship's motion at point Q will be moving slightly faster than one who participates in the ship's motion at point P, as the ship travels a curved path, because objects at the top of the mast will move with a more rapid circular motion than those on the bottom of the ship.

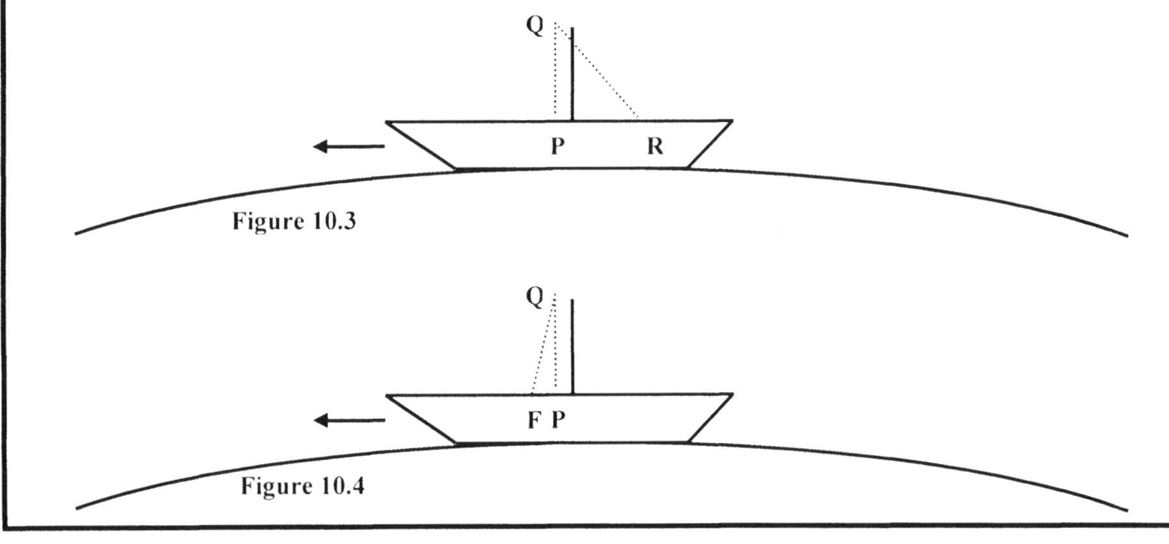

Figure 10.3

Figure 10.4

Newton on Inertia and Gravitational Attraction

While Galileo was close to the principle of inertia, it was Isaac Newton who first correctly formulated the principle: An unperturbed body in straight-line motion in empty space will stay in motion; one at rest will remain at rest. How then did Newton explain the elliptical paths of planets?

Following the diagram in Figure 10.5 below, a body at point A has an inertial tendency to travel along the path described by line AB, but receives a gravitational impulse at B (i.e., motion directed toward point S) and now has an inertial tendency to travel along the path described by line BC, and so on. One must understand that, for any line such as AB, comprised of an infinite number of points, there are an infinite number of impulses that ultimately enable the body here to move elliptically around point S.

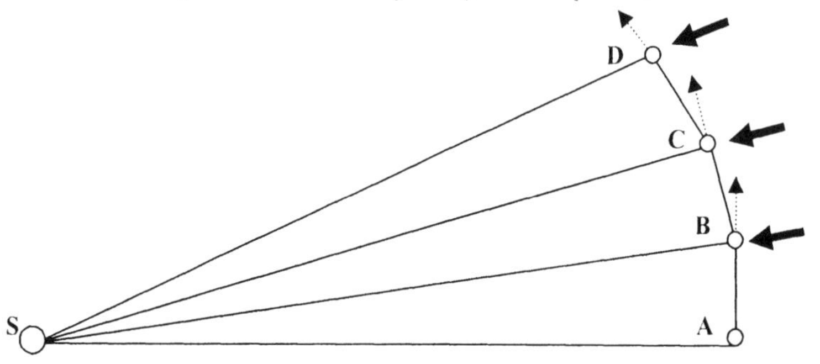

Figure 10.5 Newton on Elliptical Motion

Rene Descartes (1596-1650) has given us perhaps the most celebrated example of an axiomatic approach in his works in science and philosophy. Turning inductivism on its head, Descartes believed that no systematic approach to knowledge could get off the ground without a firm metaphysical foundation of first principles that are unquestionably true. Philosophically, beginning with the claim "I exist, whenever I think" and taking himself to have shown the existence of a supremely good and powerful deity that is not a deceiver, he went on to show, among other things, that physical bodies, as extended things, have real existence.[3] Scientifically, he offered a mechanical system that proved a suitable rival to that of Newton. In *Principles of Philosophy* (1644), Descartes stated that the universe is not a void, but a plenum, where bodies keep their motion while moving along closed loops, and that changes in bodily motion are a result of impact from other bodies.[4] He went on to deduce three other laws of motion:

LAW 1: Bodies at rest remain at rest, and bodies in motion remain in motion, unless acted upon by some other body.

LAW 2: Inertial motion is motion in a straight line.

LAW 3:

Part a: If a moving body collides with a second body, and the second body's resistance to motion exceeds the first body's force to continue its motion, then the first body changes its direction without losing any of its motion.

Part b: If the first body has greater force than the second body has resistance, then the first body carries with it the second, losing as much of its motion as it gives up to the second.

From here, he deduced seven rules governing specific types of collisions.

Descartes' picture of the cosmos—involving bodies, motion, and collision—was a kinematic alternative to the Newtonian cosmos. Though the model would prove false, it had at least a couple of advantages over Newton's model. Descartes' system involved no appeal to action at a distance and the notion of a vortex offered an explanation for why all the planets moved in the same direction.

Hume on Induction

The difficulties in trying to square deductive systems that presumably explain natural phenomena with observations of the phenomena themselves was, as we have already seen, most astutely noticed by David Hume. If science is a discipline that involves necessary connections among phenomena through universal laws that are, in some sense, driven inductively or non-demonstratively by observation, then there ought to be some justification for this method of inference based on observation. Hume in his *Treatise on Human Nature* (1739) directly addressed this difficulty. Acknowledging the existence of two types of reasoning—that involving relations of ideas (deductive reasoning) and that involving matters of fact (inductive reasoning)—Hume goes on to ask how we can justify claims involving matters of fact.

Whereas truths involving relations of ideas are necessarily true and have no bearing on empirical reality (i.e., axiomatic systems like Euclid's geometry), those of matters of fact are only true insofar as they agree with observable reality, which can always be otherwise. His inquiry takes the form of a *causal* analysis of empirical phenomena. Consider the claim, "Flame is a cause of heat". Analysis of this causal claim reveals:

> Flame and heat are *contiguous*,
> Flame is *temporally prior* to heat,
> Flame and heat are *constantly conjoined*, and
> Flame and heat are *necessarily connected.*

Of these conditions, Hume maintains that only the first three seem to be justified by experience; the last one is a conceptual condition that seems to beg the very question of the legitimacy of induction as a viable form of argument.

In short, the problem of induction is one of justifying the use of the method of reasoning that goes from what has been observed to what has been unobserved. For instance,

> All *observed* instances of flame have been hot.
> So, all instance of flame are hot.

Now one cannot fully justify inferences of this sort unless one can give some sort of assurance that the future will be like the past. No such appeal to reason is helpful, since reason must acknowledge the possibility of the future not being like the past. One can always imagine that, as of some particular time, the physical laws governing the universe might be changed, say, by a fickle deity, who has distaste for laws obtaining for more than 10,000 years. On the other hand, one can always argue that inferences from observed to unobserved have always worked in the past. Yet this is question-begging, as it is a mere *observation* that inferences from observed to unobserved have always worked. What, then, will guarantee that this new observation will work? This, in effect, would be to use an inductive argument to justify inductive reasoning.[5]

The moral of Hume's message for us today is that science itself must be taken on faith of some sort: Science only makes sense on assumption that nature is uniform—that is, that the future will be like the past—and, though we have no reason to doubt that, we cannot have any assurance of it.

Goodman's New Riddle of Induction

Hume's problem was given a different twist by Nelson Goodman[6] in the 20th century in *Fact, Fiction, and Forecast*. He has us consider the following two hypotheses:

H_1: All emeralds are green.
H_2: All emeralds are grue.

Goodman has us take "grue" to mean "green if observed before some future time, blue otherwise". Let us set this future time to be 2121. Now H_1 implies that all emeralds before and after 2121 are green. H_2, in contrast, implies that all emeralds before 2121 are green and all after 2121 will be blue. Now as both hypotheses say the same thing about the color of emeralds before 2121, each is equally well confirmed. Yet both cannot be true, since they are inconsistent. Which hypothesis is correct?

The problem can be intensified by adding a third hypothesis,

H_3: All emeralds are grurple,

where "grurple" means "green before 2121, purple thereafter". There is, as this suggests, no end to the predicates I could create—grack, gred, gravender, grorange, etc.

Moreover, there is no reason to think that "grue" and "grurple" are predicates less legitimate than green. One who sincerely predicated "grurple" of emeralds would, I suppose, see the world differently. He would fully expect all emeralds to turn purple in the year 2121. Thus, what Goodman shows is that, for any particular hypothesis, one can come up with an indefinitely large number of competing hypotheses that are confirmed equally well by observation, though inconsistent with each other.

Given Goodman's problem about the legitimacy of scientific predicates and Hume's problem about the legitimacy of inductive inference, it is not surprising to see why some philosophers of science have turned away from any hope of there being a fixed logic underlying science and turned towards giving merely descriptive accounts of the actual practice of science. Such ponderous difficulties lead us to ask the following question: Is scientific understanding possible?

Is Scientific Understanding Possible?

Problems like those of Hume and Goodman show that there are clear limits in the very practice of science. Science is, after all, human activity and thereby subject to the limits of human experience and human apprehension. In short, it is wise not to think that science holds the promise of solving the most fundamental riddles of the cosmos or even the greatest problems facing humans.

As science attempts to make sense the world around us, to make sense of science, we must begin with faith of a sort. We must take on faith the notion that natural phenomena are law-governed, for, as Hume has shown, we cannot prove this and, without this assumption, scientific practice would make little sense. Only by assuming that all things are governed by discoverable laws does it make sense to strive to discover these laws.

Of course, we need not be dismayed or even soured. Science, if anything, seems to bewilder and amaze us each day through new discoveries, creations, and helpful information. Some of these, like discoveries about deep-space objects, merely sate our curiosity about how things work. Others—like drugs to fight AIDS, ATM machines, and remote controls for appliances—may be put to use toward lengthening, simplifying, or even softening our lives, should we desire longer, simpler, or softer lives. Moreover, the very success of science seems itself some evidence for the regularity of all things. Yet this of course question-begging, is it not? Still, if the cosmos itself is regular, then science, an ordered investigation of this regularity, *must* work.[7] If it is not, then we do not seem to be any worse for having practiced science.

Key Terms

scientific a-priorism
Newton, inertia, and elliptical motion
Descartes' laws of motion
Goodman's new riddle of induction
Galileo on inertia
Cartesian foundationalism
Hume on induction

Text Questions

* What are some of the key problems with a-priori accounts of the way the world works?
* How does the problem of induction for Goodman differ from the problem as Hume elaborated it? Precisely how does each formulation bear on science today?

Text-Box Questions

* How did Galileo nearly come up with the law of inertia?
* What was Galileo's theory of tides and how was it an attempt to confirm the Copernican, sun-centered system?
* How did Newton explain planetary motion as the result of his universal law of gravity and his law of inertia?
* What were some of the difficulties that led to the break up of Aristotle's world-view in the 16th and 17th centuries? What one discipline did Descartes appeal to in his attempt to reconstruct metaphysics and a new world-view?

---

1 Newton correctly identified inertia as a moving body's tendency to remain in straight-line motion or a resting body's tendency to remain at rest in empty space.

2 From what Newton called his "rules of reasoning in philosophy". Isaac Newton, *Mathematical Principles of Natural Philosophy,* trans. A. Motte (Berkeley: University of California Press, 1962).

3 Rene Descartes, *Meditations on a First Philosophy* ed. John Cottingham (New York: Cambridge University Press, 1996).

4 Rene Descartes, *Principles of Philosophy* in *The Philosophical Writings of Descartes, Volume I*, ed. John Cottingham et al. (New York: Cambridge University Press, 1985).

5 David Hume, *A Treatise of Human Nature*, ed. David Fate Norton and Mary J. Norton (New York: Oxford University Press, 2000) & *An Enquiry concerning Human Understanding*, ed. Tom L. Beauchamp (New York: Oxford University Press, 1999).

6 Nelson Goodman, *Fact, Fiction, and Forecast* (Indianapolis: The Bobbs-Merrill Co., Inc., 1965).

7 Hans Reichenbach, "The Pragmatic Justification of Induction", *The Theory of Knowledge*, ed. Louis P. Pojman. (Boston: Wadsworth, 2003).

Module 11
Logical Accounts of Science

"If we are prepared to deny that there is any truth to our most successful theories on the grounds that it is possible for a false theory to be successful, we might as well deny that there is an external world on the grounds that it is possible for a brain in a vat to have external-worldish experiences". Jarrett Leplin, "Truth and Scientific Progress"

WHAT DIFFERENTIATES THE HYPOTHESES OF SCIENCE from hypotheses of a non-scientific nature is the public accessibility of the former. This is to say that the claims made by scientists need not be taken on faith; the results may be repeated by additional experiments or confirmed by appeal to observation. The only requirements for a scientific outlook are a commitment towards rationality and an admission of fallibility.

What guides the practice of science are *the generation of hypotheses* and *testing for their fit with reality*. To the question, "How are scientific hypotheses generated?", no one answer is possible. To give a definitive answer amounts to, I suppose, buying into one of the several different schools of thought on the question "How are scientific hypotheses justified?" These questions, concerning scientific discovery and scientific confirmation, are the focus of Modules 11 and 12.

The next two modules take a small, selective sample of some of the more influential accounts of the nature of science. I arrange these into "logical" (Module 11) and "historical" accounts of science (Module 12). The current philosophy-of-science literature, as it exists today, is admittedly substantially more complex.

This module looks at logical accounts of science. I use the term "logical" loosely to indicate by adherents of this school an essential commitment to the use of some form of logical structure in the practice of science. I begin with a sketch of early Inductivist approaches to science, turn to the Hypothetico-Deductive Method, and end with a purely deductivist approach to science—Falsificationism.

Inductivism & the Generation of Hypotheses

Beginning with Aristotle, the earliest approaches to giving a systematic account of the generation of scientific hypotheses (i.e., scientific discovery) were based on inductive arguments.

> **INDUCTIVISM is the view that inductive argument plays an important role in science, whether in the generation of hypotheses or in their confirmation.**

The use of induction as a method of generating scientific generalizations has been practiced by such notables as Aristotle, John Duns Scotus, William of Ockham, Francis Bacon, Jakob Bernoulli (1664-1705), David Hume, John Herschel, William Whewell and John Stuart Mill. Each has argued that scientific hypotheses are in some sense created by an inductive or non-demonstrative inspection of evidence that, as it were, points to the appropriate generalization. I offer the reader here a mere taste. Mill's inductive methods of causal analysis were examined in Module 8.

Perhaps the most important spokesperson of the inductivist movement was Francis Bacon in his work *Novum Organon* (1620). Bacon's work was a harbinger of a new scientific methodology—a data-driven, inductive approach to science. Taking himself to be ushering in a new age, Bacon stated that the prior study of natural phenomena had been obscured by *four idols*. First, the Idols of the Tribe concerned the defects of human nature that predisposed men to shoddy scientific practice. Second, the Idols of the Cave concerned personal attitudes toward experience that was due to faulty upbringing and education. Third, the Idols of the Marketplace concerned the common, vulgar meaning of words that was an impediment to clarity in scientific conceptions. Last, the Idols of the Theatre, a poke mostly at Aristotelian teleology, concerned biases on account of the dogmas and methods of different schools of philosophy.

Bacon insisted that science must be founded on factual observations. Thereafter, scientists should seek relationships among these facts through inductive generalization. To separate accidental relationships from those that were essential, Bacon utilized another inductive procedure, his method of exclusion, to complement inductive generalization. In such a manner, he thought, essential relationships could be separated from accidental relationships, and the former, having passed the test of exclusion, would be considered correct generalizations from facts. These generalizations or relationships would then be the subject matter or "facts" of the next, more inclu-

Aristotle on Philosophy of Science

In a proper scientific demonstration, the deduction must say more than that something is the case; it must tell why something is the case. For this, two conditions must be met. First, the middle term of the demonstrative syllogism, the term common to both premises, must be causal, not just descriptive. Second, the argument can only be laid out with universal and affirmative premises in the first figure. Aristotle sums, "If it is something else and it is demonstrable, then the cause must be a middle term, and it must be shown in the first figure, since what is being proved is universal and affirmative".

To illustrate just how to form a correct demonstration, Aristotle has us consider two deductions: one of the fact and one of the reason why. A deduction of the fact tells us merely that something is the case, but does not give us a proper explanation—one involving the cause. For instance, Aristotle has us consider a properly formed first-figure syllogism (having three universal claims: "All A are B") that shows the problem involved with deductions of the fact.

P_1: All things that do not twinkle are things that are near.
P_2: All planets are things that do not twinkle.
C: All planets are things that are near.

This suggests that planets are things that are near because they are things that do not twinkle. Yet, not twinkling is not a cause of nearness; it is nearness that causes them not to twinkle. The causal explanation is the other way around: Planets are things that do not twinkle because they are things that are near. In other words, the deduction of the fact is not demonstrative because it has the wrong middle term. In contrast, proper demonstration (the deduction of the reason why) is illustrated as follows:

P_1: All things that are near are things that do no twinkle.
P_2: All planets are things that are near.
C: All planets are things that do not twinkle.

Here, nearness is the reason why planets do not twinkle.

sive level of generalizations. At the highest level of generalization were Forms, which described the relationships between the simplest naturally existing, agentive entities.[1]

A fine example of the inductive approach to science is *Boyle's law*, which states that the volume of a gas in any closed container stands in inverse proportion to its pressure, or $p \propto 1/V$. Boyle presumably came upon this law simply by observing and recording the relationship between the pressure and the volume of a gas in containers. The data themselves then straightforwardly suggested the law.

Later thinkers developed most sophisticated account of scientific inductivism. John Herschel believed that there were *two paths to law-like generalizations*.[2] One was, in agreement with Bacon, through inductive schemata, of the sort illustrated above. A second was through bold conjecture, unaided by inductive schemata—in a word, educated hunches. In framing such a distinction, Herschel was the first to distinguish *scientific discovery* from *scientific justification*. It is one question, he thought, to ask how a generalization is generated (i.e., scientific discovery) and quite another to ask whether there is sufficient evidence to accept it as true (i.e., scientific justification). Only the latter, he believed, was entirely within the province of inductive reasoning, as the former sometimes involved scientific intuition or hunches.

Another inductivist approach to science was by William Whewell.[3] Whewell believed that the warp and woof of science were facts and ideas. At the lowest level of induction, science begins by collecting perceptions of experiences or "facts" and then groups these basic facts under ideas—rational principles or generalizations that bind together facts. As ideas tell us which facts to group together, there can be no notion of fact independent of some idea. At a higher level of generality, these ideas then serve as facts for another level of ideas at a higher level of generality.

How, for Whewell, were ideas generated? This was the very essence of induction. He believed that several tentative hypotheses ought to be considered for any group of facts. However, he did use the guidelines of simplicity and symmetry. For example, he did not consider hypothesis generation as a rule-governed enterprise. Guesswork was often needed. What proved an idea or hypothesis to be correct was its capacity to bind together certain related facts in need of explanation. If this idea *qua* law could be linked with other ideas, bound together as fact, and then tied to a more fundamental idea, then this would bring about a "*consilience of inductions*". It is not unlike observing a few meager streams meeting to form a brook and then a few brooks meeting to form a river. This consilience would offer additional inductive confirmation that the system as a whole is correct. What motivated Whewell here was the staggering success of Newton's system to accommodate Kepler's laws of planetary motion and Galileo's insights on moving bodies.

There many difficulties that face inductivists. I mention here the most vexing. As we have seen in the previous module, there is the problem of the legitimacy of induction as a form of argument that can justify general claims. For Hume, we recall, no amount of past experience can prove a theory true, as no amount of past experience can guarantee that the future will be like the past. Another way of saying this is that no amount of finite observations can justify a general claim, as the latter is potentially infinite in scope. How, then, does inductive reasoning serve to justify the law-like generalizations of science? Any attempt at justification will either be circular (i.e., will employ an inductive argument in justification of inductive reasoning) or beg the question (i.e., it will work on assumption that the universe will continue to be uniform. Why, we are left to ask, is induction preferable to, say, guessing or any other method?

For Goodman, we recall, the difficulty has a slightly different twist: Distinct conceptions of natural phenomena (e.g., "All emeralds are *green*" vs. "All emeralds are *grue*"; see Module 10) can lead to different expectations of what the future will be.

Moreover, if we concede that law-like generalizations can be generated through enumerative induction, we still cannot explain how certain laws could have come about in this manner. What observations, for example, could have prompted Galileo to formulate his law of falling bodies ($d \propto t^2$, in a vacuum)?

Lamarck on Speciation

Prior to Darwin's theory of natural selection (see Module 12), Jean Baptiste de Monet Lamarck (1744-1829) published his *Philosophie Zoologique* (1809) to explain the existence of the various species observed and the functions of their parts.

Lamarck began through asserting four chief principles. He stated, first, that there is a tendency in nature toward organic complexity and, second, that the environment acted to promote variation in organisms. He writes:

> It will become clear that the state in which we find any animal is, on the one hand, the result of the increasing complexity of organization tending to form a regular gradation; and, on the other hand, of the influence of a multitude of very various conditions ever tending to destroy the regularity in the gradation of the increasing complexity of organization.

Just how did the environment produce such variations? His answer included his third and fourth principles—the principle of use and disuse of parts, which Darwin would use as a secondary mechanism of change, and the principle of the inheritance of acquired characteristics. In other words, the continuous use of an organ gradually strengthens that organ (while continuous disuse of an organ weakens it) and what is strengthened or weakened over time tends to get passed on through sexual reproduction.

Though Lamarck's explanation of the formation of species appealed to nothing other than natural causes, seeing Nature herself as working toward some end, he could not escape the widespread teleology of his day. For him, all species were a part of a large, branched scale (a *scala naturae*) that tended toward perfection. At the bottom of the scale, where the branches are most numerous, nature spontaneously formed great numbers of the simplest, least vigorous and least perfect types of plants and animals (when heat, sunlight and electricity acted on unorganized, moist, gelatinous matter). Over time and through circumstances, others organisms—less in number, more vigorous, and most perfect—formed from these.

Lamarck recognized two types of animal: invertebrates and vertebrates. Looking only at the vertebrates, fishes in time developed into reptiles, which developed into birds, which developed into mammals. Not surprisingly, of the mammals, humans, the most perfect of all organisms, were said to be at the very top of this scale.

Invertebrate Animals		**Vertebrate Animals**
1. Molluscs	6. Insects	1. Mammals
2. Cirrhipedes	7. Worms	2. Birds
3. Annelids	8. Radiarians	3. Reptiles
4. Crustaceans	9. Polyps	4. Fishes
5. Arachinds	10. Infusorians	

Still with Lamarck, we do have a break with the notion of fixity of species assumed by Catastrophists (see Module 6). According to Lamarck, species are artificial constructions. Appearances aside, there is in reality a continual progression from lower forms of life to higher ones. We categorize by species only because local environments create local variations that make differences seem pronounced and because of gaps in our knowledge of reality.

Deductivism

Hypothetico-Deductive Account

With the difficulties facing inductivism, it is not surprising that many philosophers, wedded to there being an underlying logic to science, have adopted a deductivist approach. Broadly construed,

> **DEDUCTIVISM is the view that deduction plays a key role in hypothesis confirmation/disconfirmation or in deciding between competing hypotheses.**

The *hypothetico-deductive account*, a form of deductivism, has its roots in the work of thinkers such as Herschel, Whewell, and Isaac Newton and is proposed by some philosophers as a more reasonable alternative to pure scientific inductivism for generating hypotheses. According to the hypothetico-deductive method, no hypotheses are generated inductively, through observation; instead they are *invented* in some manner—merely thrown out for consideration. The formation of hypotheses, on this account, is a matter for psychology to investigate, not philosophy. However, once the hypothesis is put forth, a prediction is deduced from it as a test for the truth of the hypothesis. A situation is set up to test the prediction. If the test situation agrees with the prediction, the hypothesis is said to be *confirmed*—a process that makes the hypothesis more likely to be true than it was prior to the test. If the prediction turns out false, the hypothesis is said to be *disconfirmed*. Disconfirmation is a matter of rejecting the hypothesis outright, though, as we shall see immediately below, the case for outright rejection is seldom warranted, without further thought. In summary:

Confirmation	*Disconfirmation*
H → P	H → P
P	~P
So, H	So, ~H

The H-D method is one of the most attractive views of how scientific hypotheses are justified, though it has little to say about how hypotheses are generated.

One difficulty, which shows that the schematization above is overly simple, is suggested by the *Duhem-Quine thesis*.

> **According to the DUHEM-QUINE THESIS, no hypothesis is absolutely disconfirmable because there are always auxiliary hypotheses that may rescue it from refutation.**

Perhaps the best way to explain the thesis is to give an illustration of it. One of the most celebrated illustrations of the thesis in the history of science is the discovery of Neptune. A slight perturbation was noticed in the orbit of Uranus—seemingly inconsistent with Newton's law of gravity and three laws of motion. As all attempts to reconcile the inconsistency faltered, it seemed that this inconsistency was itself disconfirmatory evidence against Newton's system.

$H \rightarrow P$
$\sim P$
So, $\sim H$

(Where H = "The universe is a Newtonian system" & P = "If H is true, the orbit of Uranus will be at such-and-such position at such-and-such time")

Race for the New Planet

Given the laws of Newtonian dynamics, the measurable anomaly in the orbit of Uranus could only be accounted for by an unseen gravitational mass pulling on it. In 1845, Cambridge mathematician John Couch Adams used the laws of Newton's system to predict the existence of a hitherto undiscovered planet. Having made the calculations, he now only needed to look for the new planet.

Adams went to Sir George Airy, Astronomer Royal, of the Royal Observatory at Greenwich. As Airy was unavailable, Adams left behind a message with his calculations. Airy was slow to act on the data. It was only when French mathematician Urbain Leverrier had published a prediction similar to that of Adams did Airy undertake the project of looking for the new planet. He asked James Challis of Cambridge University to search for it. Challis began in July of 1846. Leverrier undertook his search with the aid of German astronomers at the Berlin Observatory. The race was on and Leverrier's Franco-German team beat out the British team by discovering the new planet, thereafter named Neptune, on the night of September 23 of 1846. Astonishingly, Challis had actually observed Neptune on four different occasions prior to this without recognizing it.

Astronomers were reluctant to get rid of Newtonian theory of gravity, since there was nothing to take its place and it had been trouble-free up until the perturbation of Uranus. In an attempt to salvage the theory, it was suggested that the theory could still be true, should there be some perturbing force found that is impacting Uranus' orbit. Calculations were made and it was predicted that a body of a specific mass would have to be at specific distance from Uranus to account precisely for the deviation from its projected orbit. Sure enough, Neptune was discovered. What seemed to be disconfirmatory evidence wound up being further confirmation of Newton's theory.

What this example of the Duhem-Quine thesis shows is that falsification does not refute a hypothesis alone, but a hypothesis in conjunction with certain auxiliary hypotheses ($A_1, ..., A_n$, below). Schematically:

$(H \ \& \ [A_1, ..., A_n]) \rightarrow P$
$\sim P$
So, $\sim(H \ \& \ [A_1, ..., A_n])$

Thus, the conclusion, $\sim(H \ \& \ [A_1, ..., A_n])$, does not tell us that H is definitely false, but rather that H *or* at least one of the auxiliary hypotheses is false—that is, $\sim H$ *or* $\sim(A_1, ..., A_n)$.[4]

Falsificationism

Falsificationism, another deductivist account, is the view made popular by Karl Popper in an attempt to solve the age-old problem of induction.[5] Popper argued that inductive reasoning played no role in hypothesis testing, as, in agreement with Hume, it could do nothing to give us grounds for thinking a hypothesis to be true. The true test for any scientific hypothesis, he thought, was its ability to survive rigorous tests designed to disconfirm or refute it. The best sorts of hypotheses, then, were those that offered the most numerous and the riskiest tests—thereby lending themselves open to refutation. What marks off statements of true science from those of pseudoscience, for Popper, is that they are in principle straightforwardly refutable; no *ad hoc* hypotheses can rescue them and the predictions they make are themselves exposing the hypothesis to great risk of falsification. In short, the best sorts of hypotheses are those that are clearly articulated, simple, and boldly stated—that is, they have high empirical content.

To illustrate, from a finite list of competing hypotheses for some phenomenon, $H_1,\ldots H_n$, science proceeds by eliminating those hypotheses that are untenable, until such time, if ever, that only one is left. What is left, having stood the test of time, is likely to be true through elimination. Justification of a hypothesis, then, is merely a matter of surviving all rigorous attempts at refutation.

Consider three competing hypotheses concerning the extinction of dinosaurs during the K-T period.

H_1: Mammalian egg predation resulted in the extinction of the dinosaurs.
H_2: A giant asteroid crashed into the earth's atmosphere and killed off the dinosaurs.
H_3: Epidemic disease wiped out all the dinosaurs.

First, it seems impossible that mammalian predators (H_1) could have eaten enough dinosaur eggs to cause their extinction. Again, what would have accounted for the mass extinction of so many other species of animals at the time? Finally, this hypothesis does not account for environmental changes at the time such as drop in sea level and global warming. It cannot be maintained. Therefore,

$H_1 \rightarrow P$
$\sim P$
So, $\sim H_1$

Thus, from our original list of contenders, we can eliminate one.

H_1, H_2, or H_3
$\sim H_1$
So, H_2 or H_3

Second, the epidemic-disease hypothesis (H_3) is equally flawed. It too fails to explain the environmental changes of the time. In addition, it seems absurd to think that any epidemic disease, from what we know today about how epidemic diseases behave, could wipe out so completely the population of dinosaurs. Therefore, from our remaining list of contenders, we can eliminate another.

H_2 or H_3
$\sim H_3$
So, H_2

Now we are left with one hypothesis—H_2. If this hypothesis yields specific testable predictions and *every attempt at falsifying through putting the hypothesis to the test fails*, then Popper maintains we can say that the hypothesis is "well-corroborated". Unfortunately, there are problems with H_2 as well.

The problems with falsificationism, however attractive it may seem, are many. First, how can one ever be sure that the initial group of hypotheses gathered is exhaustive—that is, that within it is the true hypothesis? If it is not, then we cannot know that the conclusion is true. In the case of dinosaur extinction, for instance, we have not entertained a number of other reasonable hypotheses. Second, surviving refutation is not the same thing as bringing forth positive evidence that some hypothesis is true, so surviving refutation seems to be a rather impoverished sense of "justification". Third, what are we to do when there are many unfalsified contenders, as is the actually case with the extinction of dinosaurs? Here complexity, ad-hocness, and designing further experiments for testing come into play.

Last, there are other problems with Popper's insistence that he has solved the problem of induction. First, to call a theory that survives all tests one that is "well-corroborated", an assessment that admits of degrees, if it is to have any meaning, seems just to be another way of saying that it is strongly confirmed—that it has good inductive support. If it does not have this meaning, and Popper asserts it does not, then the tag "well-corroborated" is simply reducible to "as-of-yet-not-falsified", which seems to be inadequate for a theory that aims at truth. Second, what does it mean to say that hypotheses with greater empirical content are preferable, prior to being put to the test, to those with lesser empirical content? To say of one hypothesis that it has greater empirical content seems, if anything, to be a claim to the effect that such a hypothesis has proven itself (i.e., escaped falsification) in the past and that it will likely do the same in the future. This seems quite like an inductive inference, in spite of what Popper says. If so, then Popper's claim to avoid all inductive reasoning is false. Finally, there is Popper's blanket statement that inductive reasoning plays no role at all in science. Yet he also asserts that risk is crucial to science (see Module 14)—i.e., that predictions that are likely to be true even if a hypothesis is false are in the main worthless in science. Assessment of risk, however, presupposes the use of some method (prior experience) to assess risk. To do this is to smuggle in induction, at least analogically, through an assessment based on prior expectations. Consider two claims that hitherto have never been put to a test: "If I drink the pint of Ol' Aged Ale when thirsty, my thirst will be slaked" and "If I drink the pint of Ol' Aged Ale when thirsty, my throat will be parched". One cannot say prior to any test, as Popper would, that the second is decidedly riskier than the former without some use of previous experience about drinking ales, which Popper forbids.

Key Terms

hypothesis generation
inductivism
Boyle's law
Herschel's two paths
deductivism
hypothesis confirmation
hypothesis justification
Bacon's four idols
Lamarck on speciation
Whewell's consilience of inductions
hypothetico-deductive method
hypothesis disconfirmation

Duhem-Quine thesis
Falsificationism
discovery of Neptune

Text Questions

* What is the difference between the generation of hypotheses and their confirmation? In your answer, list the relative strengths and limits of the inductivist and deductivist approaches listed above.
* What are Bacon's four "Idols" and how is each an impediment to the practice of science?
* How does confirmation differ from disconfirmation according to the hypothetico-deductive method?
* What is "corroboration" for Popper and how does it differ from "confirmation"?

Text-Box Questions

* What is the difference for Aristotle between a "deduction of the fact" and a "deduction of the reason why"? Which is appropriate for a scientific demonstration?
* What was Lamarck's view of speciation? Be sure to include all four principles or mechanisms of change in your answer. Why is "species" itself an artificial construction for Lamarck?

---

1 Francis Bacon, *Novum Organum*, ed. Peter Urbach & John Gibson (Open Court Publishing Company, 1994).
2 John F. W. Herschel, *A Preliminary Discourse on the Study of Natural Philosophy* (London: Longman etc., 1830).
3 William Whewell, *Philosophy of the Inductive Sciences* (London: John W. Parker, 1847) & *Novum Organon Renovatum* (London: John W. Parker & Son, 1858).
4 See Pierre Duhem, *the Aim and Structure of Physical Theory* (New York: Atheneum, 1962) & Willard van Orman Quine, "Two Dogmas of Empiricism" in *From a Logical Point of Veiw* (Cambridge: Harvard University Press, 1953.)
5 Karl Popper, *The Logic of Scientific Discovery* (New York: Harper & Row, Publishers, 1968) & *Conjectures and Refutations* (New York: Basic Books, 1963).

Module 12
Historical Accounts of Science

"In the course of the centuries the naïve self-love of men has had to submit to two major blows at the hands of science. The first was when they learned that our earth was not the center of the universe but only a tiny fragment of a cosmic system of scarcely imaginable vastness. This is associated in our minds with the name of Copernicus,.... The second blow fell when biological research destroyed man's supposedly privileged place in creation and proved his descent from the animal kingdom and his ineradicable animal nature. This revaluation has been accomplished in our own days by Darwin, Wallace and their predecessors, though not without the most violent contemporary opposition. But human megalomania will have suffered its third and most wounding blow from the psychological research of the present time which seeks to prove to the ego that it is not even master in its own house, but must content itself with scanty information of what is going on unconsciously in its mind". Freud, *Introductory Lectures*, XVIII.

CRITICS CHARGE THAT PHILOSOPHERS OF SCIENCE in the logical tradition, in their zeal to disclose the nature of the scientific method, have failed in one significant way: Their descriptions of the method of science fail to map on to the practice of science. Thus, their accounts are too restrictive.

This module examines history-of-science approaches to science. Again, use of the word "history" does not imply that adherents of this trend completely disregard reasoning in science, only that adherents believe that the actual practice of science is not completely governed by immutable rules of reasoning. History-of-science accounts do, however, mark a trend away from the normativism (an ideal at which scientists aim or, at least, ought to aim) of logical accounts toward descriptivism (an account of what scientists in fact do) as a philosophical ideal.

I sketch out below three of the most influential accounts in the last several decades.

Paradigm Relativism: Normal and Revolutionary Science

One of the difficulties with logical approaches to scientific hypotheses and scientific progress is with disconfirmation. If anything, what the actual practice of science clearly shows is that a theory that is false, but useful, does not necessarily get thrown into the junkyard of falsified theories.

Again let us return to Newtonian mechanics. In spite of its success in helping astronomers discovery Neptune, there were unresolved difficulties with the theory, such as a slight perturbation in the orbit of Mercury, some four arc-seconds in its perihelion, for which Newtonianism could not account. As was the case with the perturbation of Uranus' orbit, scientists tried to save Newtonian theory by positing a small planet internal to the orbit of Mercury called "Vulcan" that was causing the perturbation. Although no such planet was ever found, Newton's mechanical views were not formally rejected until the formulation and subsequent successes of the General

Theory of Relativity, centuries later. Nonetheless, Newton's "laws" are still today used for sending out satellites and space ships.

Cartesian Foundationalism

The Copernican heliocentric model of the universe, published in the year of Copernicus' death (1543), offered a challenge to the Ptolemaic/Aristotelian model of the universe. However, the work was not widely read—perhaps due to its removal of the earth from the privileged position at the center of the universe—and its impact would not be felt till adherents after Copernicus, Galileo most notably, could provide more substantive arguments on its behalf. Moreover, Copernicus' model was not a substantial threat to Aristotelian physics, as it relied heavily on many of Aristotle's principles of physical science.

By the early part of the 17th century, numerous problems had surfaced for the Aristotelian world-view, which, we recall (see Module 3), was separated into a superlunary realm and sublunary realm. Precise measurements by Tycho Brahe of strange and unexpected astronomical phenomena, like the supernova of 1572 and the comet of 1577, showed that, in spite of the prohibitions of Aristotle, change did occur in the superlunary realm of the cosmos. Moreover, Galileo's observations of four moons orbiting Jupiter violated the Aristotelian dictum that all heavenly motions were motions around the earth. These observations were additional difficulties to those already acknowledged to exist in the sublunary realm for Aristotle's model of the cosmos. For instance, Aristotle's notion that bodies fall in a medium (e.g., air or water) in proportion to their weight was demonstrably false. So too was Aristotle's account of projectile motion, which his physics could not cleanly explain. With the Aristotelian world-view crumbling, by the start of the 18th century, the philosophical and scientific communities were in a state of crisis.

With the foundations of just about everything that was believed to be true being now called into question, scholars scurried to find anything that was beyond doubt in an effort to obtain a suitable ground for a new approach to philosophical and scientific understanding. For one scholar, Rene Descartes, this foundation was to be had in the only science that seemed beyond reproach: geometry. So it was to Euclid that he turned—specifically, the Euclid's deductivism. Unlike the inductivism of Bacon, Descartes promoted an a priori approach to the founding of first principles—one, he thought, so rationally compelling that these first principles could not be reasonably doubted. He cast into doubt all objects of sensory experience and even the truths of mathematics, until he came to the one proposition that, upon scrutiny, could not be doubted. He writes in his *Meditations*: "'I am, I exist' is necessarily true every time I utter it or conceive it in my mind". From this rather meager beginning, Descartes then took himself to have proved the existence of god and the existence of external objects, whose essence is extension, as well as certain natural, mechanical laws about the way these objects interact. This approach disregarded any appeal to experience—at least in establishing the most fundamental truths of metaphysics and the most fundamental laws of the physical world.

That a theory with problems ought not always to be rejected was stated by Thomas Kuhn, one of the most notable critics of logical approaches to science (see "Scientific Relativism", Module 6). In his *Structure of Scientific Revolutions*,[1] Kuhn argues that science is not progressive, but problem-solving. As the problems of one time or culture are not the same as those of another, so too do the problems of science differ within different scientific communities over time. Problems, then, generate scientific practice, which defies description through any sort of immutable logical analysis.

The basic unit of scientific analysis for Kuhn is the *paradigm*. A paradigm, also called an "accepted model or pattern",[2] is roughly understood as the basic framework of a particular scientific community at a given time. It is a way of seeing the world that generates, though it is not defined by, its own methodologies and rules. Within a scientific community, there are periods of *normal science* in which a paradigm, like Ptolemaic geocentrism, becomes established and remains unchallenged by rival paradigms. Problems or anomalies emerge in time. Yet no single anomaly is sufficient to pose a significant threat to a paradigm. When enough anomalies surface to challenge a dominant paradigm, a *scientific revolution* ensues. Kuhn writes, "[S]cientific revolutions are … those non-cumulative developmental episodes in which an older paradigm is replaced in whole or in part by an incompatible new one".[3] A new paradigm, having its own language and rules and being "incommensurable" with the old one, replaces the old one, "whole or in part", in an effort to solve the problems the old paradigm could not solve. Scientists now *see* and *live in* a different world. Problem-solving ability, however, is not the same as a paradigm's truth or falsity—something entirely outside of the aims of science for Kuhn. Thus, there is nothing in principle disallowing an abandoned paradigm from being adopted later, as a paradigm justifies itself exclusively by its ability to settle current problems.[4]

Thus, for Kuhn and other like-minded historicists, the error of logical approaches comes precisely in their refusal to examine closely the practice of science in their "discovery" of the logic of science. There is for Kuhn no set-for-all-time logic of science, as this logic is itself endemic to a particular paradigm and can change with a paradigm shift.

There is considerable merit in Kuhn's insistence that philosophical accounts of science square with the practice of science. He is also right in stressing that much of scientific practice has an irrational component. Still, what seems most troubling for Kuhn is that he maintains science is not in the business of seeking truth—not even in limited capacity. One wants to say, or instance, that adoption of Galileo's law of falling bodies, $d \propto t^2$, is not merely a matter of a paradigm shift from Aristotelian science, which claims that bodies fall in proportion to their weight in some medium. Aristotle's view is manifestly wrong *in any paradigm*. Failure to recognize that makes the practice of science relativistic. Moreover, Kuhn states that different paradigms are not only inconsistent, but are "often actually incommensurable".[5] That is at least vague, if not false. If, by this, he means that the language of one paradigm has no or few points of contact with the language with another, then that seems completely overstated and flatly incorrect. That the Copernican world-view is incommensurable with Ptolemy's world-view is manifestly false: Both theories work under the assumption that Aristotle's physics is true. If he means that the complete system of propositions of, say, the General Theory of Relativity is a different system and a different way of looking at the world than that of Newtonian Dynamics, then that is, I believe, trivially true. That is certainly not the same as incommensurability and that is certainly is not to say that those in different paradigms see different things. They see the same things, but interpret what they see in a different manner.[6] That Kuhn's work has had such appeal, given the problems with his view, is probably very much due to the shortcomings of the logical school of thought and not to the merits of his own view.

Research-Programmes Approach

Critically following the work of Kuhn, Imre Lakatos proposed that a critique of science begin not with paradigms, but research programmes.[7] For Lakatos a *research programme* consists of a hard core of irrefutable propositions, a protective belt of auxiliary hypotheses, and a heuristic for solving problems that arise. The heuristic is constituted by methodological rules—research paths to avoid and follow that function to protect the hard core. Paths to avoid are called the "negative heuristic" of a research programme; paths to follow, the "positive heuristic".

Using Newtonian mechanics as an example, let us revisit the problem of the perturbation of the orbit of Uranus. For Lakatos, it is absurd to attack the hard core—the law of gravity, the three laws of motion, and other laws that are part of the Newtonian system (Kepler's laws of planetary motion, Galileo's law of falling bodies, etc.)—so the heuristic suggests the least damaging alternative. When the situation is examined fully, it is quite clear that the predicted path of Uranus is contingent on there being no other orbiting body that perturbs its motion. Once this assumption or auxiliary hypothesis is called into question, there is the possibility of salvaging the hard core by rejecting this auxiliary hypothesis. Scientists did just this and, upon examination, they discovered Neptune and gained further confirmatory evidence of Newtonian dynamics, just when it looked as if there was reason to reject Newton's dynamical system.

Isaac Newton (1643-1727)

Isaac Newton is commonly regarded as on of the greatest physicist, mathematicians, and astronomers of all time. In his *Principia Mathematica,* he discovered the law of universal gravitation and the three laws of motion (see Module 5), and thereby unified all bodies in the universe, both terrestrial and celestial, under the same theoretical framework. In developing this system, Newton developed calculus, independently of Gottfried Leibniz, and also articulated the laws of conservation of momentum and conservation of angular momentum. The deterministic system that Newton advanced was indispensible for establishing Copernicanism, and it had powerful scientific appeal, until Einstein's work on relativity theory in the early 20th century.

In addition, Newton's work in Optics marked breakthroughs in the science of his day. He put forth a corpuscular theory of light, and observed that white light, passing through a prism, comprised a spectrum of colors.

In spite of his remarkable scientific achievements, Newton devoted the lion's share of his time to the study of alchemy.

According to Lakatos, the positive heuristic is the hard core of principles of a programme that are deemed by convention untouchable or irrefutable. During the course of a research programme, there forms a protective covering of auxiliary propositions that function ever increasing to secure the core of irrefutable principles by buffering them from refutation. This, of course, is not to say that this hard core will never be overturned. The history of science clearly shows otherwise. Still, in such a manner, seemingly disconfirmatory evidence does not necessarily tell against the research programme as a whole, but instead may force reconsideration of some of the auxiliary hypotheses. In extraordinary circumstances, where an anomaly cannot be reconciled with the research programme and no more comprehensive theory currently exists to explain the anomaly, the only alternative might be to shelve the anomaly until such time that it can be explained.

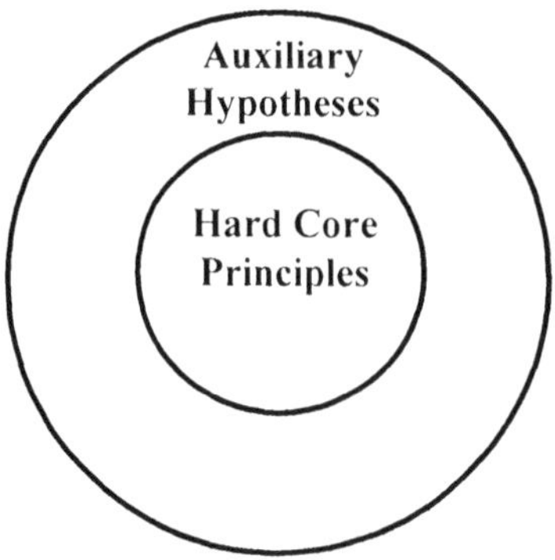

Figure 12.1 Lakatos' Research Programme

Overall, the advantage of Lakatos' approach is that it tries to reconcile a historical approach of science with one that is methodological or logical. He also tries to salvage the notion of scientific progress. For Lakatos, any particular series of theories is *progressive* when these three conditions are met:

> C_1: The most recent theory in the series explains all that the previous theories explained,
> C_2: The most recent theory has greater empirical content than any of the others, and
> C_3: Some of this excess empirical content has been corroborated scientifically.[8]

Lakatos did admit that research programmes could become sterile and degenerative. The mark of a healthy research programme is its capacity to predict and accommodate new information.

A difficulty for this approach is that, though chiefly historical, it uses *incorporation* as a measure of progress, which is a form of coherence. If research programme's are historically driven, nothing in the research programme is ultimately immune from rejection, and every research programme is assessed by its ability to incorporate a previous programme, then incorporation itself must be decided by the methodological rules of the previous research programme, which is being rejected. If not, there must be an appeal to some neutral methodological system that stands outside of the history of science to decide the issue, which Lakatos does not allow. Thus, to say of research programme R_2 that it is superior to research programme R_1 is to say that, by the methodological rules of R_1, R_2 has shown itself as superior. This certainly does not seem to be a reasonable means of salvaging progress.

Progress and Problem Solving

Another historical approach that has attracted much attention over the years is that of Larry Laudan. In his influential work, *Progress and its Problems*, Laudan defines science as a problem-solving enterprise that is progressive:

> The rationale for accepting or rejecting any theory is thus fundamentally based on the idea of *problem-solving progress*. If one research tradition has solved more important problems than its rivals, then accepting that tradition is rational precisely to the degree that we are aiming to "progress," i.e., to maximize the scope of solved problems. In other words, the choice of one tradition over its rivals is a progressive (and thus a rational) choice precisely to the extent that the chosen tradition is a better problem solver than its rivals.[9]

Thus, in contrast to logical approaches, which argue for an independently existing standard of rationality and state that scientific theories are progressive insofar as they meet this standard, for Laudan, scientific progress (i.e., efficiency in solving problems) determines scientific rationality.

The problems of which Laudan speaks are both conceptual and empirical and both work within a particular domain of science. *Conceptual problems* occur at the theoretical and methodological levels of science, as when two inconsistent rival theories, like the biological theories of Lamarck and Darwin on speciation, are put forth to explain data. *Empirical problems* concern

Darwin on the Origin of Species

In 1859, Charles Darwin (1809-1882) published *The Origin of Species*—the culmination of years of research during his observations on the H.M.S. Beagle (1831-1836). In this work, Darwin challenged some of the traditional principles of biology: the fixity of species and teleology (the notion that nature does nothing in vain, see Module 3). Prior to its publication, his views on natural selection were given a lukewarm reception.

The key to *natural selection* is a principle Darwin claims to have got from Thomas Malthus: As more individuals of a species are born than can survive, there is in consequence a struggle for existence. In this struggle, only those best fitted to survive, those most suited to their environment, will likely survive.

> In the preservation of favored individuals and races, during the constantly recurrent Struggle for Existence, we see the most powerful and ever-acting means of selection. More individuals are born than can survive. A grain in the balance will determine which individual shall live and which shall die—which variety or species shall increase in number, and which shall decrease, or finally become extinct.

Nature, then, acts as a selective force that, as it were, picks out those best suited for survival. "I have called this principle, by which each slight variation, if useful, is preserved, by the term Natural Selection, in order to mark its relation to man's power of selection".

To get natural selection off the ground, Darwin assumed the existence of biological variability in living things. In other words, living things within a particular species vary, howsoever slightly, from individual to individual, and variability is in general biologically adaptive. Moreover, these variations are passed on to offspring.

Yet natural selection, thought the driving force in evolution, was not the sole mechanism of change. Like Lamarck, he believed that use/disuse played a role. "Use in our domestic animals has strengthened and enlarged certain parts, and disuse diminished them; and that such modifications are inherited".

By positing that nature was a selective force, Darwin was not speaking teleologically—as if nature were some sort of selecting agent that acts for the best of each organism—as were Aristotle and Lamarck. Instead, Darwin was describing a statistical tendency in nature. Those organisms that were best adapted to their environment would have a greater chance of survival and passing on their traits. Lions that are strong and fast will have a greater likelihood of catching gazelles. Likewise, gazelles that are slow and weak will have a greater likelihood of perishing and passing on their traits. As such, in a stable environment, there will be tendency for organisms to progress toward an ideal, determined by structural limits.

Nature, Malthus taught Darwin, is prodigal, not frugal, and prodigality is not teleological.

> As natural selection acts solely by accumulating slight, successive favorable variations, it can produce no great or sudden modification; it can act only by very short and slow steps. Hence the canon of "Natura non facit saltum" [Nature does not make jumps], which every fresh addition to our knowledge tends to make truer, is on this theory simply intelligible. We can see plainly why nature is prodigal in variety, though niggard in innovation. But why this should be a law of nature if each species has been independently created, no man can explain.

Thus, for Darwin, nature has no need of supernature to explain speciation. Through the transfer of slight variations over very long periods of time, entirely new species form.

domain objects, as when trying to account for the existence of "satellites" orbiting Jupiter with an astrophysics that is Aristotelian.

Given this framework, Laudan maintains that progress may occur in one of a few ways:

C_1) Progress occurs when there is a reconciliation of two seemingly conflicting theories,
C_2) Progress occurs when any anomaly is resolved (e.g., the discovery of Neptune), or
C_3) Progress occurs when there is an increase in the number of empirical problems solved.

The difficulty here, as with Lakatos' approach, is that there is no independent appeal to a fixed notion of rationality to decide progress. The criteria of rationality are fixed by the criteria of progress. Why, then, should anyone adopt this view of progress?

He attempts to solve this difficulty by an appeal to the *scientific elite* of a particular time and *standard-case episodes* of science. Different methodologies are assessed by their capacity to reconstruct the largest number of standard-case episodes. The assessors are the scientific elite, and their judgments are held to be indisputable. Still, what counts as a standard-case episode must be determined by the scientific elite and this seems entirely arbitrary without a fixed methodological rules—that is, some form of accepted logic. There is a vicious circularity here. One must have a methodology to decide that something is a standard-case episode, but the standard-case episodes decide the best methodology. Laudon's account of progress, then, is unavailing.

In a later work, *Science and Values*, Laudan has put forth his *reticulational model*[10] in an effort to escape some of the difficulties of his early view. Here methodological principles (M), theories (T), and axiological aims (i.e., values) (A) are each involved in scientific justification. Unlike the Kuhnian model, where a change of paradigms involves a change in each of these elements, for Laudan, the reticulational model allows for gradual changes in a scientific community. Change just means that there has been some adjustment among M, A, or T—or some combination of the three in more radical cases. The benefit of this approach, unlike that of Kuhn, is that it allows for subtle changes in a scientific community. It also allows for the possibility of changes in the methodological assessment in science. Laudan concedes that his reticulational model does not escape relativism altogether, but merely the type of holistic relativism of Kuhn—characterized by the *complete* paradigm shift.

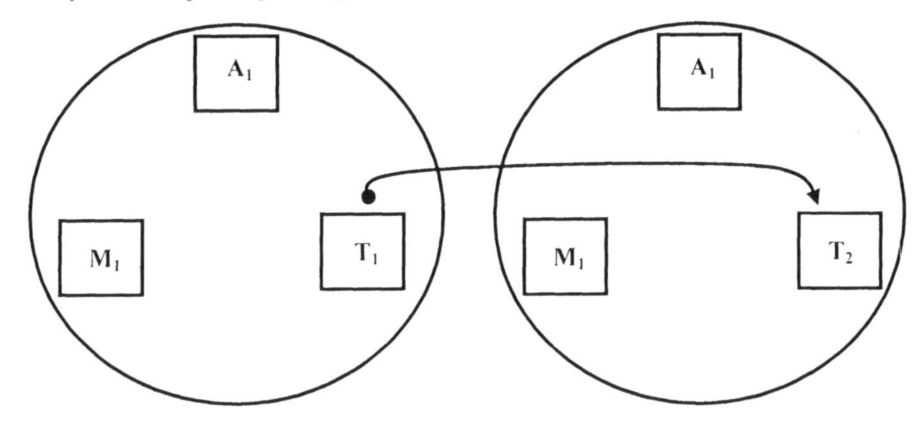

Figure 12.2 Laudan on Paradigm Change. For Laudan, paradigm change occurs whenever any one or any combination of the basic constituents of a paradigm (M, A, or T) changes. Here there is a paradigm change at the theoretical level, as T_1 changes to T_2. Kuhnian paradigm shifts, considered complete shifts, would be a change of M, A, and T.

Key Terms

paradigm
revolutionary science
research programme
normal science
Cartesian foundationalism
progressive theories

SECTION FIVE

Scientific Models

Module 13
Three Types of Scientific Hypotheses

"I know that I am mortal and the creature of a day; but when I search out the massed wheeling circles of the stars, my feet no longer touch the earth, but, side by side with Zeus himself, I take my fill of ambrosia, the food of the gods". Claudius Ptolemy

OVER 2000 YEARS AGO, Aristotle said that *bodies fell to the earth* simply because *all heavy objects have a natural tendency to fall to the center of the earth.* Though this was believed true by most scholars for centuries, eventually the notion that bodies do what they do just because they have a natural tendency to do what they do (i.e., teleological explanation) came to be seen as empty and *ad hoc*. Moliere satirically mocked Aristotle's teleology in *Le Malade Imaginaire*, where in a medieval doctoral exam, the doctoral candidate explained why opium puts people to sleep by stating to the learned doctors that it contained a dormative principle. Centuries after Aristotle, Newton gave a different explanation of why bodies fall to the earth. He stated that bodies fall to the earth because *there is an attractive force between all bodies with mass.* In short, a stone dropped to the earth falls because the great mass of the earth tugs on the negligible mass of the stone and draws the stone to it. To deflect anticipated criticism of how bodies that are not in contact with each other can act on each other through the mysterious force of attraction, we recall Newton's famous reply, "I frame no hypotheses". Finally, in the early part of the 20th century, Einstein proposed a radically different and descriptively more accurate explanation of why bodies fall to the earth. He said that *all massive objects warp the space around them (in effect, creating an accelerative field)* and, thus, bodies fall to the earth because *this is the easiest path for them to take in warped space.* In short, for Einstein, space was not the empty, featureless "container" that Newton said it was; it was a sort of empty thing that had at least one feature—it could be warped.

Each of the three attempts to explain why bodies fall to the earth as they do is a scientific hypothesis. Roughly speaking, a scientific hypothesis, as we mention in Module 4, is an educated guess about some phenomenon in need of explanation.

A HYPOTHESIS is a statement about the nature of some part of reality in an effort to gain a fuller understanding of that part of reality.

Its scope can be very broad (e.g., The universe is expanding at an ever-accelerating rate) or very narrow (e.g., Summertime in Thessaloniki is always very hot). In a word, a hypothesis attempts to discover order or regularity in the world of observable phenomena. Ideally, each hypothesis generates at least one prediction that is straightforwardly testable. Passing the test of predictive consequences is a measure of the success of a hypothesis. The more rigorous the test to which a prediction puts a hypothesis is, the greater the likelihood that the hypothesis is true upon passing such a test. Upon passing test after test, it becomes less likely that the hypothesis is false and more likely that what the hypothesis says about reality is true of reality.

Hypotheses are the workhorses of scientific investigation. Without them, no scientific work would ever begin. In what follows, I introduce three types of scientific hypotheses: universal hypotheses, statistical hypotheses, and causal hypotheses.

Three Types of Scientific Hypothesis

Universal Hypotheses

Following the lead of Aristotle (see Module 3), scientific hypotheses properly take the form of universal statements. Let us use for illustration Einstein's Theory of Relativity, which has two fundamental principles that can be framed as universal hypotheses (i.e., having the statement form "All α are β"):

UH_1: All (α) inertial frames are (β) reference frames in which the most basic laws of physics are the same.
UH_2: All (α) inertial frames are (β) reference frames in which the speed of light is the same.

One consequence of the Theory of Relativity is that the speed of light—roughly 300,000 kilometers per second (186,000 miles per second)—is an upper limit to all material objects in the universe. As a material object gets closer to this limit, it takes an increasingly greater amount of energy to accelerate it. Thus, it would take an infinite amount of energy to move any massive object to the speed of light.

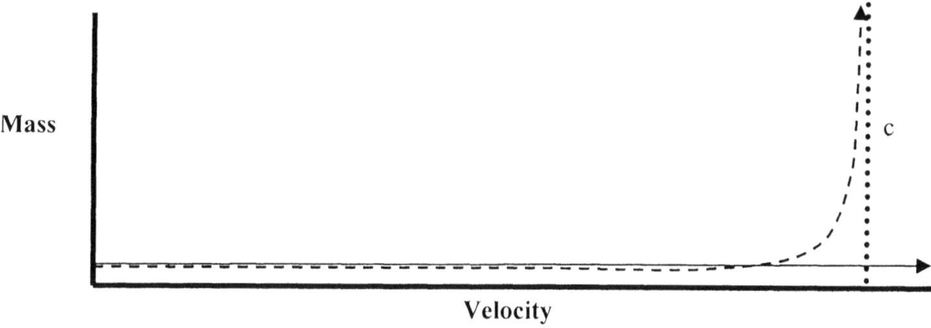

Figure 13.1 $E = mc^2$ vs. $F = ma$. The dotted line represents the acceleration of any massive object in an Einsteinian universe ($E = mc^2$, or energy equals mass times the speed of light squared: dashed arrow). As it nears the speed of light, c (an upper limit for all objects), it requires more and more energy to move it. Since c is a constant here, the increasing energy cannot be straightforwardly transferred to its velocity, so it is transferred to the object's mass. The body, gaining tremendously in mass, moves asymptotically toward c as its limit. In a Newtonian universe ($F = ma$, or force equals mass time acceleration: full line), there is no such limit and the body's mass is a constant. Here, simply, increasing force translates into increasing acceleration.

That the speed of light is an upper limit for all massive objects according to the Special Theory of Relativity makes for an interesting contrast between the Einsteinian universe and that of Newton. Imagine an individual in a train that is traveling at a constant speed of 70 kilometers per hour. He points a strong flashlight in the direction in which the train is traveling and shoots out a beam of light. How fast is the light traveling? According to Einstein, the answer is 300,000 kilometers per second, as measured by the person on the train *as well as* by a curious observer, who stands near the tracks, but is not on the train. In short, there is no privileged position! According to Newton, however, the beam of light should be measured to be 300,000 kilometers per second by the one on the train, but 300,000 kilometers per second *plus* 70 kilometers per hour to the person who merely watches the train.[1]

In short, for Newton, velocities are additive with respect to the canvas of absolute space; for Einstein, velocities are not simply additive. For instance, imagine someone riding a bicycle at 20 miles per hour who throws forward a ball at 30 miles per hour. To an observer standing idly by, according to Newton, the ball would be traveling at 50 miles per hour (20 + 30). According to Einstein, with the speed of light as an upper limit to all motion, the ball would be measured at a speed just slightly less than 50 miles per hour ($v_1 + v_2/[1 + (v_1v_2/c^2)]$) or ($20 + 30/[1 + (20 \times 30/c^2)]$), (where c = the speed of light in a vacuum)—that is, roughly 49.9996 miles per hour. These consequences, however paradoxical, have been put to the test and shown to be correct.

The Twin Paradox

What Relativity Theory shows is that both time and space are not absolutes, as Newton maintained, but relative to one's vantage point. And this has been shown to have some rather paradoxical consequences. Image a set of twins: Roger and Ranger. Roger is placed in a rocket ship that travels out in space and then back toward earth at a speed that is 0.9998 that of light. When Roger returns to earth, he notes that he has spent one year in space. When he lands, much to his astonishment, everything has changed remarkably. Most notably, Ranger has aged 50 years in one year's time! From Ranger's perspective, Roger has spent 50 years in space but has aged only one year.

As strange as this might seem, experiments have confirmed the relativity of space and time. In 1971, cesium atomic clocks were flown on commercial airliners around the world in both directions. Their time was then compared to the time elapsed on an earthbound clock. Their eastbound clock lost 59 nanoseconds; their westbound clock gained 273 nanoseconds. These measurements agree with predictions of the Special Theory of Relativity.

Statistical Hypotheses

Determinism, as we have seen briefly in Module 7, is the view that *every event has a sufficient cause*. In other words, whatever happens at some point in time is completely determined by prior conditions of the universe as well as the laws of the universe. Determinism is a working hypothesis of scientists who work at the macroscopic level of phenomena. The presumed regularity of observable phenomena presupposes that determinism is true.

If determinism is true at the macro-level of things—that is, if a complete causal explanation is in principle possible for any phenomenon—then why is there a need of statistical hypotheses (i.e., having the statement form "X% of α are β") in science? The answer is simple: Since it is impossible to know all the causal factors that go into a particular event—since we cannot have a god's-eye perspective on events—statistical hypotheses are a measure of our uncertainty about them. Thus, statistical hypotheses are of great practical value in many sciences—like sociology, psychology and medicine—where large amounts of data are collected and stored and there is some attempt to make sense of all these data.

Consider the following two statistical hypotheses:

SH_1: (α) Smoking is strongly correlated with (β) lung cancer.

SH_2: (α) People who regularly attend church are (β) people with a 25% reduction in mortality.

SH_1 is a statistical claim (though no percentages are given) that has the support of decades of solid research. SH_2, in contrast, comes from relatively recent research conducted on religion and bodily health that has had some startling preliminary results, in spite of problems with the research methods.

The differences between these two statistical hypotheses notwithstanding, there are important reasons why we use statistical hypotheses in a macro-world that is assumed to be entirely deterministic.

First, as SH_1 and SH_2 suggest, *statistical hypotheses can be useful guides to decision making.* Let us consider a young, high-school athlete, Filbert Buckley, who wants to become a professional football player. A statistically relevant question for him would be: What is the likelihood of a high-school football player making it to the ranks of professionalism? Here, I suppose, it is merely a matter of a simple proportion. Filbert could take the overall number of high-school football players from the last five years (say, HS) and place this in proportion to the number of those football players who made it as professional players (say, P)—that is P/HS. He might find that he has a one-in-6,000 chance of making it to the professional ranks. That result, no doubt, would be disheartening to him, especially if Filbert is not among the elite high-school athletes.

Moreover, *statistical hypotheses generate causally relevant information.* Gretchen Meurington, who has recently suffered a stroke, may wish to examine food-science studies in an effort to use improved diet as a means to improved health. For her, studies that suggest certain foods are correlated with cardiac health could be important sources of causally relevant information that could change her consumption patterns. For instance, on finding some evidence that dark beer prohibits blood clotting, Gretchen may decide to add one bottle of imperial stout to her daily diet.

This, of course, leads us to another type of hypothesis, linked both to universal and statistical hypotheses—the causal hypothesis.

Causal Hypotheses

Causal hypotheses (i.e., having the statement form "α is causally related to β") differ from those that are universal or statistical in that they imply a special relationship between the categories involved in them. Let us consider the two statistical hypotheses, considered above, reformulated as causal hypotheses:

CH_1: (α) Smoking causes (β) lung cancer.
CH_2: (α) Regular religious worship is causally related to (β) longevity.

In both of these hypotheses, there is a correlation between each of the variables considered. CH_1 is a hypothesis where the correlation, established by decades of research, is quite strong—so strong that it is taken as a *matter of fact* that smoking causes lung cancer. This, of course, does not mean that anyone who smokes at all will develop lung cancer. It merely means that heavy smoking is a predictor of lung cancer in later life. While this, for many, is tremendously informative, it is, for others unhelpfully vague. On the one hand, many who have never smoked have gotten lung cancer. On the other, many heavy smokers have lived long lives without getting lung cancer.

The issue, of course, is complex not only because the human body is complex, but also because so many external factors, other than smoking, affect the human body in a multiplicity of ways. Even in the tightest of laboratory conditions, where researchers can control for external variables in the best possible manner, it is impossible to control for all conceivably relevant variables. Thus, causal claims are virtually impossible to establish as true. The best we can do is to speak with some degree of certainty that some factor, α, is the cause of another, β. In effect, in establishing a causal relationship between variables, we really do no more than show a tight and undeniable correlation—one so tight (may Hume forgive us!) that the evidence suggests a necessary connection between the two. In contrast to CH_1, when we look back to CH_2 we find a causal claim without such a tight connection between categories. As a causal claim without much backing, it is best construed as a statistical hypothesis—at least, until such time as additional, unambiguous data can be brought forth to establish a causal connection. I shall say more about causation and causal hypotheses in later sections. For now, it is enough to show that causal hypotheses may be construed as statistical claims with a bit more umph….

KEY TERMS

hypothesis
statistical hypothesis
universal hypothesis
causal hypothesis

TEXT QUESTIONS

* What are hypotheses and why are the called the "workhorses of scientific investigation"? How many types of hypotheses are there?
* Given that the speed of light, *c*, is a constant for Einstein, what happens to a body whose velocity increases over time? In short, why is Einstein forced to replace $F = ma$ with $E = mc^2$?

TEXT-BOX QUESTION

How does confirmation of the twin paradox show that space and time are not absolutes?

---

1 The real privileged position for Newton was absolute space.

Module 14
Theoretical Models, Hypotheses, & Theories

"Man is but a reed, the most feeble thing in nature; but he is a thinking reed. The entire universe need not arm itself to crush him. A vapor, a drop of water suffices to kill him. But, if the universe were to crush him, man would still be more noble than that which killed him because he knows that he dies and the advantage which the universe has over him; the universe knows nothing of this". Blaise Pascal, *Pensees*

Models & Theories

What is it that scientists do? Scientists, in part, construct models of reality. These models may take various forms. James Watson and Francis Crick, in constructing the double-helix model of DNA, set up the structure by seeing DNA sequencing to be like a spiraling staircase. Physicist J.J. Thomson gave an early model of the atom based upon an analogy with plum pudding; Ernest Rutherford, later, modeled the atom after a planetary system. Other models are made to scale, as when someone, wanting a visual grasp of how our solar system works, might try to build an exact replica of it on a much smaller scale.

The most fundamental and most precise model used by scientists is the theoretical model. A theoretical model is a propositional description of some aspect of reality.

A THEORETICAL MODEL is a consistent propositional system that attempts to explain fully some aspect of reality through at least one theoretical hypothesis.

What links models to reality, through predictions that they make, are theoretical hypotheses. They are, as it were, the nuts and bolts of theoretical science. Since theoretical hypotheses almost always take the form of universal statements, I offer the following definition.

A THEORETICAL HYPOTHESIS is a statement of the form "All α are β" that is part of the propositional network of at least one theoretical model.

Here the term in subject position, α, is universally distributed; not so with β. In other words, the statement is saying something about each and every thing that is a member of category α—namely, that every member of category α is also a member of category β; while the converse need not to hold (except in cases where α and β are identical sets). For instance, "All penguins are animals incapable of flight" asserts that each and every *penguin* belongs to the category *flightless animals*, but it does not assert that each and every *flightless animal* belongs to the category *penguin*. Below are three illustrations:

Darwin on Speciation (Two Claims):

1. All (α) living organisms are (β) entities that have evolved over the course of hundreds of thousands of years from simpler organisms.

2. All (α) change from one species into another is (β) change that occurs gradually and persistently over hundreds of thousands of years.

Freud's Model of Dreams (Two Claims):
1. All (α) *dreams* are (β) *phenomena generated by unconscious childhood wishes.*
2. All (α) *dreams* are (β) *phenomena fulfilling those wishes.*

Newton's Law of Gravity ($F=Mm/r^2$):
All (α) *instances of attraction of any two mass points* are (β) *forces equal to the mass of the first body and the mass of the second body over the square of the distance between them.*

Scientific Laws

Scientists use the term "law" in a variety of ways. For example, laws can express relationships that are:

DISPOSITIONAL (e.g., Carnot's Law, or 2nd Law of Thermodynamics: All naturally occurring processes in an isolated system are naturally occurring processes that move in the direction of increased entropy),
IDEALIZED (e.g., Law of Falling Bodies: All distances of a body falling in a vacuum are distances that are in proportion to the square of the time),
CONSTITUTIONAL (e.g., All emeralds are green), and
STATISTICAL (e.g., All times that are the half-life of $carbon_{14}$ are times equivalent to 5760 years).

Moreover, the propositional content of laws can sometimes be expressed numerically (e.g., $E = mc^2$) and always qualitatively (e.g., Organisms evolve over time in response to changes in their environment).

However, many philosophers—such as Fred Dretske and David Armstrong—assert that if the notion of "law" is to carry any weight scientifically, it must express more than just a contingent connection between phenomena. Consider, for instance, the following two propositions:

P_1: All post-WW II U.S. pennies are copper.
P_2: All emeralds are green.

The first is a universal claim that is true, but only contingently so. There is nothing to prevent U.S. mints from make pennies out of another common metal. The second, in contrast, expresses a necessary relationship between something being an emerald and that thing also being green. Thus, a law implies a necessary connection between things and a necessary connection can handle counterfactual claims (i.e., "*Were* the wax heated to such-and-such temperature, it *would have* melted"). Yet it is impossible to establish a necessary connection between phenomena through experience alone, as any amount of data collected on such a connection is finite and a necessary connection relates to unobserved phenomena of like kind as well. Thus, a distinction between laws and regular patterns of events is presupposed in the very practice of science.

Attempts to justify certain statements as law-like, while avoiding the difficulty of trying to establish a necessary connection between phenomena, have been many and varied. Some philosophers skirt the issue by denying that laws are propositional in nature. C.J. Ducasse, for instance, argues that necessary connections can literally be observed. Mary Hesse, in contrast, believes the scope of law-like propositions is finite, not infinite.

In an attempt to show how nomological universal statements differ from those universals that are accidentally true, Ernest Nagel states that law-like universals (1) cannot be vacuously true (as is the case with "All Martians are green", since no Martians exist), (2) have predicates that are augmentable (whereas "Everything now on my plate is cold" does not), (3) are not spatially and temporally restricted, and (4) may often be confirmed by other laws in the same deductive system.

These difficulties notwithstanding, one thing is clear: The notion of law plays an important role in science. Without laws, scientific theories could not get off the ground.

In the first example, most experts on evolutionary biology accept Darwin's first generalization flatly. The second claim, however, is known to be false. There are instances where we literally "see" evolution occurring very quickly, as in mutations of the AIDS virus in host patients. In the second example, Freud himself came in time to reject the second generalization, though he had always clung to the first claim (likely also to be false). Last, Newton's Law of Gravity is a universal claim that was so well confirmed at one time that it was deemed to be a law of nature. It is still serviceable today.

For something to be a *law of nature*, if laws truly exist, is a matter of some dispute (see text box, opposite). Roughly speaking, a law describes a necessary relationship between the two categories in a universal generalization and it implies that this relationship holds in the real world. Suffice it to say, every scientific law had a rather inauspicious beginning, at some time or another, as a universal hypothesis, hungrily seeking confirmatory evidence and nomological—that is, law-like—status.

Finally, a scientific theory is a group of models linked together by common theoretical hypotheses.

A SCIENTIFIC THEORY is a consistent propositional system that attempts to explain fully some aspect of reality through certain fundamental theoretical hypotheses and a set of theoretical models that are consistent with these hypotheses.

Freudian Psychodynamics

Let me exemplify the interrelationship of theory, model, and theoretical hypothesis by means of Freud's influential psycho-dynamical theory. Methodologically, Freud began with certain assumptions that provided some background to his overall theory. I give two below, though more certainly exist:

Methodological Assumptions (Auxiliary Hypotheses):

A_1: Psychical Determinism: All psychical phenomena are events that are causally determined—that is, completely explicable by causal laws.

A_2: Recapitulationism: All individuals, from birth to old age, develop (approximately) along the same developmental path as their species does over time.

These two principles are hypotheses that are universal in scope. They are best understood as auxiliary parts of Freud's theoretical framework. Beyond this framework, Freud developed and maintained many theoretical principles—I list seven of them below—that enabled him to link up many seemingly unrelated empirical phenomena—such as jokes, dreams, parapraxes (slips of the tongue), and neurosis—each of which may be taken as a separate model, united by common theoretical principles. Freudian theory may be roughly summed below.

Theoretical Hypotheses:

H_1. All individuals are principally psycho-sexual beings.

H_2. All normal sexual development for each person follows a fixed, five-stage pattern: oral, anal, phallic, latency, and genital.

H_3. All persons have a drive to preserve life and a drive to destroy life.
H_4. Everyone's Unconscious is a storehouse of childhood sexual fantasies and phylogenetic memories.
H_5. All Oedipal resolution is needed for proper human sexual development.
H_6: All psycho-sexual tension builds and needs periodic release for proper human functioning.
H_7: All humans are constitutionally bisexual.

Turning now to his model of dreams, there are two fundamental theoretical hypotheses that Freud put forth in his *Interpretation of Dreams.* They are:

H_{G1}. All dreams are phenomena *generated* by unconscious childhood wishes.
H_{F2}. All dreams are phenomena *fulfilling* those wishes.

These two principles are, in a word, just reformulations of H_4 and H_6 to fit his theory of dreams. Fleshing out the three other models listed would show similar dependence on the theoretical hypotheses listed.

Popper on Risky Predictions

In a highly influential work in philosophy of science, *Conjectures and Refutations: The Growth of Scientific Knowledge*, Karl Popper (see Module 11) addresses the issue of demarcation: "When should a theory be ranked scientific?" Here he also voices his discontent with, among other things, Freudian psychodynamics. What bothered Popper was just the seeming explanatory power of the theory—namely, that all conceivable human behaviors were explicable in terms of Freud's theory. Such theories seemed omnipotent, because of their infallibility. Of such theories, he writes, "It was precisely this fact—that they always fitted, that they were always confirmed—which in the eyes of their admirers constituted the strongest argument in favor of these theories".

Popper was led to conclude that only theories that generate in advance *risky* predictions—those that seem more likely to be false than true—count toward confirmation. He gives as an example the observations of Eddington in 1919 that offered a test to Einstein's theory of relativity (see Module 1). Had Eddington not measured the bending of light during the solar eclipse to be consistent with relativity theory, Einstein's theory would probably not have gained acceptance. In short, Einstein's theory made certain risky predictions that were quite capable of proving the theory false. In contrast, it seems difficult to assert just what would count as an instance of falsification for Freudian psychodynamics. He sums:

> And as for Freud's epic of the Ego, Super-ego, and the Id, no substantially stronger claim to scientific status can be made for it than for Homer's collected stories from Olympus. These theories describe some facts, but in the manner of myths. They contain most interesting psychological suggestions, but not in a testable form.

Of course, Popper's views of Freud are not shared by all psychologists and philosophers of science. Many note that what Freud was trying to accomplish was something hitherto unheard of: the reconstruction of psychology as a science that includes psychical causes, operative in a non-metaphorical sense. In attempting to establish such a science, unlike Newton, Freud did not have a Kepler or Galileo to pave the way for him. He thus can be forgiven if the theory he articulated does not stand the test of time, in all of its details. What needs to be done, supporters believe, is to see whether the theory can be salvaged and, if so, what modifications need to be made to salvage it.

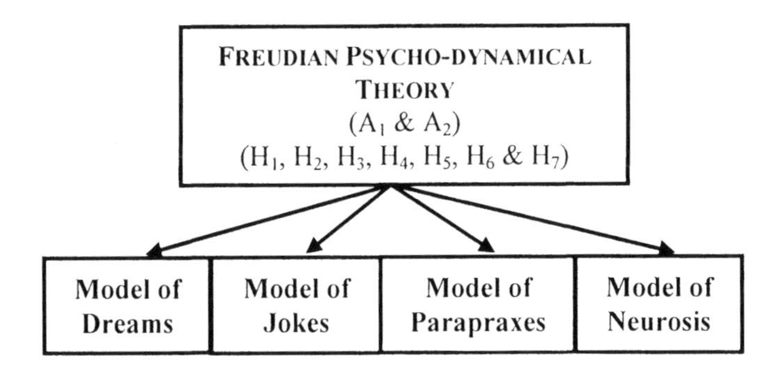

Figure 14.1 Freud's Psycho-Dynamical Theory. The theoretical hypotheses link together the various models and they also function to link the models with reality.

KEY TERMS

theoretical model
scientific theory
Freudian psychodynamics
recapitulationism
theoretical hypothesis
scientific law
psychical determinism
Popper on risky predictions

TEXT QUESTIONS

* What is a scientific theory? What role do models and hypotheses play in a theory?
* Come up with two different hypotheses to explain why some students perform better than others in school. Are these hypotheses consistent (i.e., can they both be true and part of the overall explanatory picture)? Are they inconsistent? What predictions do these hypotheses yield as tests? If the hypotheses are inconsistent, which is preferable and why?

TEXT-BOX QUESTIONS

* What are the key differences between scientific laws and universal generalizations that are not scientific laws?
* Why does Karl Popper believe that the only viable scientific hypotheses are those that generate risky predictions? How does testability factor into his criticism?

Module 15
Evaluating Models

"Even if there is only one possible unified theory, it is just a set of rules and equations. What is it that breathes fire into the equations and makes a universe for them to describe? The usual approach of science of constructing a mathematical model cannot answer the questions of why there should be a universe for the model to describe. Why does the universe go to all the bother of existing?" Stephen Hawking, *A Brief History of Time*

Some General Guidelines for Evaluation

Below, I list a few general guidelines for use in evaluating all scientific episodes—whether concerning models or statistical or causal hypotheses.

INSTITUTION/JOURNAL: *When experimental results come from a reputable institution and are published in a reputable scientific journal, never assume, before a careful analysis of the results and experimental design, that the conclusions are false or that the experimenters did not know what they were doing.*

TERMINOLOGY: *If there are technical terms relating to a report you do not understand, use a dictionary or web resources to get clear on them.*

MODALITY: *Read through the entire study/episode carefully. Look particularly for assessments of data that contain claims of possibility ("Perhaps", "It may be that", "It is not impossible that", etc.). These are introduced to weaken the claim that follows, usually because of insufficient or ambiguous data. Assess accordingly (for more, see textbox, below).*

Guidelines Specific to Scientific Models

When evaluating scientific episode of a model or theory, ask the following questions:

MODEL/THEORETICAL HYPOTHESES: *What is the model under investigation? What theoretical hypotheses drive this model?*

PREDICTION: *What prediction(s) does the model through any of its theoretical hypotheses make? The prediction should straightforwardly suggest the test that is to be done.*

BACKGROUND INFORMATION/RESEARCH METHODS: *What auxiliary hypotheses here come into play? Is there relevant background information? Is there anything worth mentioning about the research methods?*

DATA (RESULTS): *What are the relevant data? Be sure to be exhaustive here.*

ANALYSIS: *Are the data consistent with the prediction?*

If so, to what extent do they confirm it? Are there other models or hypotheses consistent with the data that cannot be ruled out?

If not, must the model be rejected categorically or is there some non-ad-hoc way to salvage it?

CONCLUSION: Give your verdict of the model here succinctly and precisely.

Statement Modality

A statement's modality relates to the manner in which it is expressed. Consider the following three sentences:

S: Big-Bang cosmology is false.
S_P: *It is possible that* Big-Bang cosmology is false.
S_N: *It is necessary that* Big-Bang cosmology is false.

From a strict logical perspective, S makes a definitive claim about a particular cosmological theory, given to explain why galaxies are moving away from each other. As Big-Bang cosmology probably has more adherents than detractors, S is bold, as it asserts that something most experts believe to be true is false. Next, S_P merely adds the modality of possibility to S and thereby weakens S substantially, since to say that any claim is possible is just to state that there is no contradiction in assuming it to be true. In effect, this is to assert very little indeed. Last, S_N asserts that S must be true. This, in effect, is to assert too much—that S is necessarily true—which, since S is not a logical truth, is not possible.

As you may have already noticed, statement modality is an important part of the assessment of scientific data. So, I shall discuss further how each of the three types of modality relates to scientific assertions, generated from research.

POSSIBILITY CLAIMS: There are two main roles for possibility claims in science.

First, possibility claims are particularly significant when someone makes a claim that goes against what is generally held to be true. Consider the following argument: "Studies show that consumption of moderate amounts of alcohol is strongly linked with prevention of heart disease. So consumption of moderate amounts of alcohol can be beneficial for health". For decades, consumption of alcohol in any amount was assumed to be harmful to one's health. These days, there are numerous studies that indicate that small to moderate amounts of alcohol consumed on a daily basis have beneficial consequences. Thus, the qualifier "can be beneficial" is appropriate in the conclusion.

Second, possibility claims are almost always suitable when data are persuasive but not conclusive, as when pilot studies are done. A pilot study is preliminary research in an effort to establish the possibility of a link between two variables, such as consumption of alcohol and improved health above. Since the suspected link between variables in pilot studies is, as it were, up for grabs, such studies usually are cost efficient. For instance, pilot studies seldom involve many subjects. Consequently, even when a correlation between variables appears, researchers are seldom in a situation, due to limited data, to draw bold conclusions. Therefore, the modality of possibility is needed for such studies.

In the main, possibility claims are a clue that data are insufficient or ambiguous and should be interpreted with due caution.

NECESSITY CLAIMS: As science is a data-driven practice, scientific claims about reality must always be tempered by the available data. This means that scientists will never be in a position to state that some claim, even the most seemingly incontestable statement (e.g., "All humans are mortal" or "Gravity is an attractive force"), is necessarily true. This holds as well for claims about reality that are said to be false, such as "Gravity is a repulsive force", since, as Hume has shown (see Module 7), it is always possible that the future will not be like the past. Strictly speaking, necessity claims are never warranted in science.

CLAIMS INVOLVING NEITHER POSSIBILITY NOR NECESSITY: Finally, conclusions from data that are not couched in terms of necessity or possibility should be read in light of the available evidence. For example, the claim "Smoking causes cancer", though very bold, is reasonable in light of the superabundance of data linking smoking with cancer. On the other hand, "Vitamin e, taken synthetically, provides significant antioxidant protection" is a dubious claim, given the ambiguity of the data on synthetic vitamins today.

Using Freud's model of dreams from the previous module as a guide, when helpful, let us now examine fully each of these six elements in turn.

Model/Theoretical Hypotheses

First, list as economically and precisely as possible the model being examined in the following form: The model M of *x* (where *x* is the phenomenon to be explained and M is the name of the model). List also all theoretical hypotheses that are parts of this model. For episodes in a newspaper or journal, the title is generally very valuable here. Going back to Freud's model of dreams, for instance:

M: The *wish-fulfillment model* (M) of dreams (*x*).
  H_{WF1}. All dreams are phenomena generated by unconscious (childhood) wishes.
  H_{WF2}. All dreams are phenomena fulfilling those wishes.

Prediction

Next, list any predictions that serve as a test for the model.

A PREDICTION is any deductive consequence of any theoretical hypothesis of a model that suggests certain experiments to test for the truth or falsity of that model.

The simplest way to formulate the prediction is to put it into the consequent of an if-then statement, with the model or theoretical hypothesis in the antecedent—i.e., "If model M (or hypothesis H) is true, then prediction P follows". Again, following Freud's wish-fulfillment model of dreams for example:

If (H_{WF2}) all dreams are phenomena fulfilling unconscious (childhood) wishes, then (P) *the recurring nightmares of former soldiers are fulfillments of unconscious (childhood) wishes.*

Background Information & Research Methods

List all relevant background information available. In general, *background information* tells readers about the motivation for this particular model. It gives them information about prior work that was done with similar aims, a list of certain auxiliary hypotheses (*include any auxiliary hypotheses here*) that drive the research, and other such things. Overall, background information enables you to decide precisely what the theoretical hypothesis is that is being put to the test.

Next, what *research methods* were used? If observations were made:

* Were these observations direct or indirect (e.g., through the use of some instrument)?
* Is there reason to believe that the observations are flawed?

If instruments were used:

* Is there reason to believe that the results of such instruments should be questioned?

If experiments were performed:

* What types of experiments were they?
* Is there reason to believe that the experiments were flawed?

Do not confuse background information and research methods with data. Data come *after* the prediction is put to the test; background information and research methods come *before* the data, generated by the prediction.

For Freud:

Freud principally used case studies from his own patients. Two auxiliary hypotheses were:

A_1: Psychical Determinism: All psychical phenomena are events that are causally determined—that is, completely explicable by causal laws.

A_2: Recapitulationism: All individuals, from birth to old age, have a developmental path that is (approximately) the same as the developmental path their species over time.

Data

The *data* are all the results of the research that have a bearing on the theoretical model's truth or falsity. Before analyzing whether the prediction turned out true or false and commenting on how well the data support or fail to support the model, make an *exhaustive list* of all the relevant data. *Leave nothing out!* You may be surprised at how the most seemingly insignificant detail can make a substantial difference in the final analysis. Remember, data are factual information as a result of observations or experiments on behalf of the model. They are not to be confused with background information and research methods.

For Freud, there are confirmatory data in abundance (see his *Interpretation of Dreams)*. Nevertheless, the recurring dreams of traumatized patients, like former soldiers, became problematic and difficult to square with H_{WF2}.

Analysis

The *analysis* is your opportunity to analyze critically the research methods, background information, and the data in order to arrive at an assessment of the model.

First, consider whether this is a pioneering attempt at modeling some aspect of reality or this is an attempt to replace an existing model by a new and better model. The background information should be helpful here.

* If this is a pioneering attempt at modeling and the data, generally quite limited for pioneering models, support the theoretical hypothesis quite well, then there is good reason to be optimistic that further research will yield similar results. Still, additional research will be needed, as preliminary research is rarely conclusive. If the data are inconsistent with the theoretical hypothesis, this in itself may provide reason enough to reject the pioneering model.
* If this is not a pioneering attempt at modeling (i.e., if the research is attempting to gather additional confirmatory evidence for an existing model) and the data support the theoretical hy-

pothesis well, then the researchers must be clear about (1) how this new research differs from and improves upon past research on this model and (2) why this new research shows that the model is superior to other models that attempt to explain the same things. If the data yield inconsistent results with the model, then something needs to be said about what may have gone wrong with this research or what may be wrong with the model. For instance, Freud's wish-fulfillment model of dreams is an attempt to replace existing physiological models of dreams with a purely psychological model. In this sense, it is remarkably novel, though strictly not pioneering.

Second, ask whether the data agree with the prediction? *If the answer is* yes, *here are some questions to consider:*

* Is the agreement superficial due to vagueness or ambiguity in the model/theoretical hypothesis or the data? For instance, is Freud's use "wish" so malleable that any conceivable dream can be construed to be the fulfillment of a wish?
* Is the prediction likely to be true even if the model/theoretical hypothesis is false?
* Are there alternative models/hypotheses consistent with the data? If so, are there any conceivable tests to decide which of these models/theoretical hypotheses is correct? Do any of these alternative models/theoretical hypotheses seem preferable for any reason (e.g., simplicity or scope; see Module 25 and Criteria of Adequacy)?
* To what extent does this model suggest further research—that is, generate new predictions through its theoretical hypotheses?
* Is the model/theoretical hypothesis consistent with generally held scientific beliefs of the time?

If the answer is no, *here are some questions to consider:*

* Are there any problems with the research (e.g., bad experimental design or faulty scientific instrumentation) that could have contaminated the results? For instance, does Freud generalize from the analysis of too many of his own dreams and not enough of the dreams of others?
* Are there auxiliary hypotheses (perhaps methodological assumptions) that should be rejected instead of the model/theoretical hypothesis?

Let us return to Freud's model of dreams. Freud came to believe that the deductive consequence of H_{WF2} was inconsistent with reality, and so he modified his wish-fulfillment thesis. In his "Revision of the Theory of Dreams" (1933), he states that recurring dreams of traumatized patients, like former soldiers, are not the fulfillment of unconscious wishes, but failed attempts at such fulfillment. Rather these traumatic dreams fulfill the wishes of the super-ego, as punishing agency. He writes:

> We should not, I think, be afraid to admit that here the function of the dream has failed.... But no doubt the exception does not overturn the rule.... You can say nevertheless that a dream is an attempt at the fulfillment of a wish. In certain circumstances a dream is only able to put its intention into effect very incompletely, or must abandon it entirely. Uncon-

> scious fixation to a trauma seems to be foremost among these obstacles to the function of dreaming.[1]

In short, Freud came to reject H_{WF2}, but not H_{WF1}.

Conclusion

Last, give your final verdict on the research as a whole. This is your *conclusion* concerning the research. As your analysis section will list all the reasons for your verdict, you need merely to give your conclusion here simply, clearly, and concisely. One sentence is often sufficient, but keep your decision as close to the data as possible. Also, be very careful in tempering your comments the scientist(s) own analysis of their model. Err on the side of caution. For instance, "Freud's model of dreams is untenable" may be too strong of a conclusion in light of the particular data available today. What seems preferable is this:

> "Freud's model of dreams is poorly supported by data and seems unlikely to be true" seems preferable.

I turn now to three illustrations of this six-stage method of analysis of models.

Mapping the Universe

Hershel's Garden Model of the Universe

William Herschel (1738-1822) of Bath was probably the most renowned astronomer of the 18th century. With the help of his sister Caroline (1750-1848), he would sweep the skies each night with home-made telescopes, the largest and best in the world at the time (7', 10', 20' and 40' refractors), in an effort to catalogue and map the various observable celestial objects. In 1781, Herschel discovered the planet Uranus. News of this discovery spread to King George III, who rewarded Herschel amply with remuneration and a prestigious royal appointment. The discoveries did not stop here however. Throughout his lifetime, he and Caroline were responsible for the discovery of numerous nebulae, clusters of stars and binary stars. In addition, he was also the first person to describe correctly the shape and size of the Milky Way Galaxy.

Throughout much of his observational career, Herschel worked under the assumption that the celestial phenomena in the sky were like specimens in a garden that develop, mature, and decay over time. This developmental model was founded upon two theoretical hypotheses: all visible celestial phenomena, however nebulous, were composed of clusters of stars and the brightness of a star was a measure of its nearness. These principles, of course, made the universe out to be a relatively uncomplicated system. They also afforded, in principle, simple tests of the model.

Herschel came to adopt these principles through simple enumerative induction. When he had come across nebulosity in the sky, he generally found that a telescope of sufficient size and resolution would resolve this nebulosity into clusters of stars.

Herschel also thought that the relative age of a cluster could be inferred by its compression and by the number of stars in the cluster (the more compressed, the older). Scattered stars came together through mutual gravitational attraction to form clusters. As many stars clustered together, they would form what merely appeared to be nebulosity.

Hitherto, all data were consistent with Hershel's Garden Model. Yet on November 13 of 1790, he writes in his journal:

> *Nov. 13, 1790. A most singular Phaenomenon! A star of about the 8th magnitude, with a faint luminous atmosphere, of circular form, and of about 3' in diameter. The star is perfectly in the center, and the atmosphere is so diluted, faint and equal throughout, that there can be no surmise of its consisting in stars; nor can there be a doubt of the evident connection between the atmosphere and the star.... [W]e therefore either have a central body which is not a star, or have a star which is involved in a shining fluid, of a nature totally unknown to us.*[2]

Now let us examine this episode in a manner consistent with the guidelines sketched above.

Search for Planet Vulcan

Module 11 tells the story of how an observational anomaly in the orbit of Uranus (discovered by Herschel), inconsistent with Newton's laws of physics, led to the discovery of Neptune. What seemed to be a reason for rejecting Newtonian dynamics thus turned out to be further confirmatory evidence of it.

In similar fashion, there was also observed in the orbit of Mercury a very small perturbation. Mercury's perihelion (closest position to the sun during orbit) was seen to advance by 43 arcseconds per century and this was pronounced to be another inconsistency with Newton's physics and Kepler's laws of planetary motion.

Years after his computations that led to the spectacular discovery of Neptune in 1846, Urbain Le Verrier predicted the existence of a planet internal to the orbit of Mercury, called "Vulcan", to account for the anomalous 43 arcseconds. The race was on to find this planet.

After decades of searching and numerous specious spottings of Vulcan, many astronomers started to doubt Vulcan's existence. The non-existence of Vulcan seemed itself an enigma, until Einstein's General Theory of Relativity was put forth in 1915. Relativity theory explained fully the anomalous 43 arcseconds as a result of the Sun warping the space around it and thereby impacting Mercury's orbit by just this amount. According to Newton, gravity is an attractive force between bodily masses and has no impact on the space around it (see Module 1, "General Theory of Relativity Confirmed"). Consequently, these two different theories each had slightly different predictions for the path of Mercury over time.

Analysis of Herschel's Garden Model of the Universe

Model/Theoretical Hypotheses

M: The *garden model* of the universe: The universe, like a garden, has specimens that develop, mature and decay in time in keeping with the following principles.

H_{GM1}: All visible celestial phenomena consist of clusters of stars.

H_{GM2}: All stars have a brightness that corresponds to its nearness to us (i.e., all stars are roughly the same size).

H_{GM3}: All clusters of stars have a relative age that is determinable by the number of stars in the cluster and the compression of those stars.

Prediction

If the Garden model of the universe is true, then *the phenomenon of November 13 is resolvable into one near star and a cluster of stars very distantly (and very directly) behind it.*

Background Information/Research Methods

* Herschel works within the Newtonian framework (i.e., he adopted the law of gravity, the three laws of motion and all other laws, like those of Kepler and Galileo, integrated into Newton's system). Note that H_{GM3} is a consequence of Newtonian Principles.
* The model shows regard for simplicity in the composition of the heavens, which, hitherto, observation has borne out.
* Herschel has built the largest and best telescopes in the world: 7', 10', 20' and 40' refractor telescopes that enabled him to see deep-space objects with sufficient resolution.
* Herschel and Caroline worked untiringly to catalogue and map out the heavens.

Data

The observed data of November 13 of 1790 seem impossible to square with the model Herschel gives. The nebulosity appears too uniform and the star appears perfectly in the center of the nebulosity so as not to be somehow connected with the star.

Analysis

If we are to take Herschel's own testimony as true—and there is no good reason to doubt his observational skills—the data seem to allow for no other interpretation than the one Herschel himself gives on November 13 of 1790. If so, in this particular case, he is forced presumably to give up the first theoretical hypothesis. The other two hypotheses must now be called into question. The model must be substantially revised or junked. (Herschel himself junks it.)

Conclusion

The Garden Model of the universe is untenable.

A Czechoslovakian Monk Studies Hybridization

Mendel on Transmission of Traits

Gregor Mendel (1822-1884) was a monk in a Czechoslovakian monastery. In a short monograph, "Experiments with Plant Hybrids",[3] Mendel was the first person to describe accurately how traits are inherited, though he could give no underlying mechanisms. So ahead of his time was his work that it took 34 years for it to be put forth independently.

Mendel's monastic lifestyle was perhaps ideal in that he had time enough to carry out the painstaking research on hybridization that was needed for his discoveries on heritable traits. Of the difficulties involved, he wrote, "That no generally applicable law of the formation and development of hybrids has yet been successfully formulated can hardly astonish anyone who is acquainted with the extent of the task and who can appreciate the difficulties with which experiments of this kind have to contend".

Experimenting with only vigorous forms of a species of garden pea (Pisum sativum) to ensure viable results, he observed seven pairs of "traits" of different pure-bred plants.

* *Shape of ripe seeds (rough and circular or wrinkled)*
* *Color of seed albumen (yellow and orange or green)*
* *Color of seed coat (white or gray-brown, with or without violet spots)*
* *Shape of ripe pods (smoothly arched and not constricted or deeply constricted and wrinkled)*
* *Color of unripe pods (green or bright yellow)*
* *Position of flowers (along the stem or at the end of the stem)*
* *Stem length (long or short)*

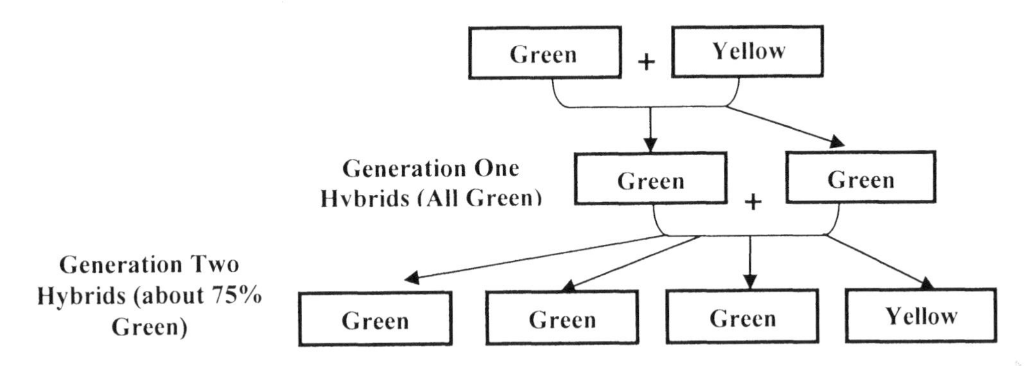

Figure 15.1 Mendel's Initial Observations on Pea-Pod Hybrids

He writes, "It was the purpose of the experiment to observe these chances for each pair of differing traits, and to deduce the law according to which they appear in successive generations". Work was done on each pair of traits. What he noticed was the following. When he cross-fertilized one pure-bred plant with another for a specific trait, say color of unripe pods (green and bright yellow), one of the traits would completely disappear (here yellow). Then, when he cross-fertilized these hybrids, the trait that disappeared in the first generation of hybrids, would reappear in the second. For instance, yellow pods reappeared in the second generation with a frequency of roughly 25 percent (see Figure 15.1, above). The data he came up with was as follows:

* *Seed shape: 7324 seeds, 2.96:1 (rough and circular: wrinkled)*
* *Albumen coloration: 8023 seeds, 3.01:1 (yellow and orange or green)*
* *Color of seed coat: 929 plants, 3.15:1 (white or gray-brown, with or without violet spots)*
* *Shape of pods: 1181 plants, 2.95:1 (smoothly arched and not constricted or deeply constricted and wrinkled)*

* *Color of unripe pods: 580, 2.82:1 (green or bright yellow)*
* *Position of flowers: 858, 3.14:1 (along the stem or at the end of the stem)*
* *Stem length: 1064, 2.98:1(long or short)*

Of the disappearance of the one factor in the first generation and its mysterious reappearance in the second generation, he writes:

> *Each of the seven hybrid traits either resembles so closely one of the two parental traits that the other escapes detection, or is so similar to it that no certain distinction can be made. This is of great importance to the definition and classification of the forms in which the offspring of hybrids appear. In the following discussion those traits that pass into hybrid association entirely or almost entirely unchanged, thus themselves representing the traits of the hybrid, are termed 'dominating,' and those that become latent in the association, 'recessive.' The trait 'recessive' was chosen because the traits so designated recede or disappear entirely in the hybrids, but reappear unchanged in their progeny....*

In an effort to explain why one factor completely disappeared in the first generation and reappeared with roughly a 25% frequency in the second generation, Mendel came up with his dominant-recessive or two-factor model of transmission of traits. His thinking was that each pure-bred plant has two "factors" for each trait—some factors being dominant; others, recessive. When a dominant factor links up with one that is recessive, during pollination, each parent plant contributes one factor from each trait in a manner that is random and the plant will only show those features of the dominant factor (Figure 15.2).

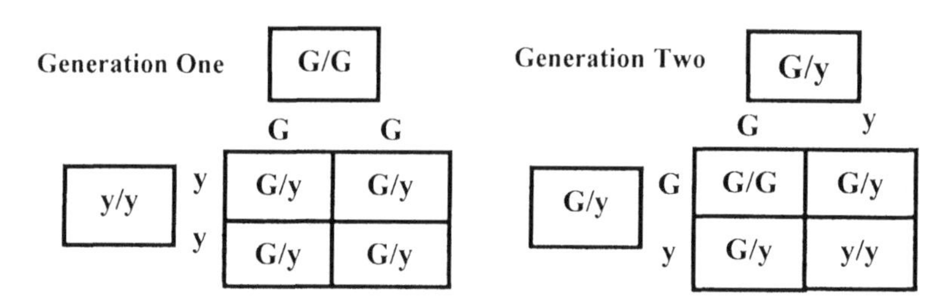

Figure 15.2. Hybrids: Generation One and Two

Analysis of Mendel on Transmission of Traits

Model/Theoretical Hypotheses

M: The *dominant-recessive factor model* of transmission of traits. For each trait (e.g., color of seed coat), there are two factors—one of which is dominant, the other of which is recessive.

H_{DRFM1}: During pollination, each parent plant contributes one factor from each trait in a manner that is random.

H_{DRFM2}: Only one factor can be exhibited in a plant and this applies to plants that actually have two unlike factors. In other words, there are two factors for each trait—one, dominant; the other, recessive.

Prediction

If the hypothesis is correct, then (1) *generation one should all be hybrids that exhibit the dominant trait, and* (2) *generation two should have half of their offspring as hybrids and half as true-breeds.* (3) *Of the generation-two true breeds, half of these will exhibit the recessive trait.*

Background Information/Research Methods

* Mendel's work what not done in true-to-form hypothetico-deductive fashion. Mendel cross bred different pea plants, gathered his data, and *then* looked at the data in an effort to come up with a model that fit the data. According to the hypothetico-deductive pattern, one comes up with a hypothesis, deduces from it a prediction and gathers data in an effort to confirm the hypothesis. This, of course, does not invalidate Mendel's theoretical hypothesis; it's only to say that he did not come about his results in standard hypothetico-deduction fashion.
* To ensure viable results, Mendel chose only vigorous plants.

Data

In crossing two true-breed plants, Mendel saw that in the first generation of hybridization one of the factors completely disappeared in all offspring. Yet when crossing these hybrids together, the factor that disappeared would reappear, though only with roughly a 25% probability.

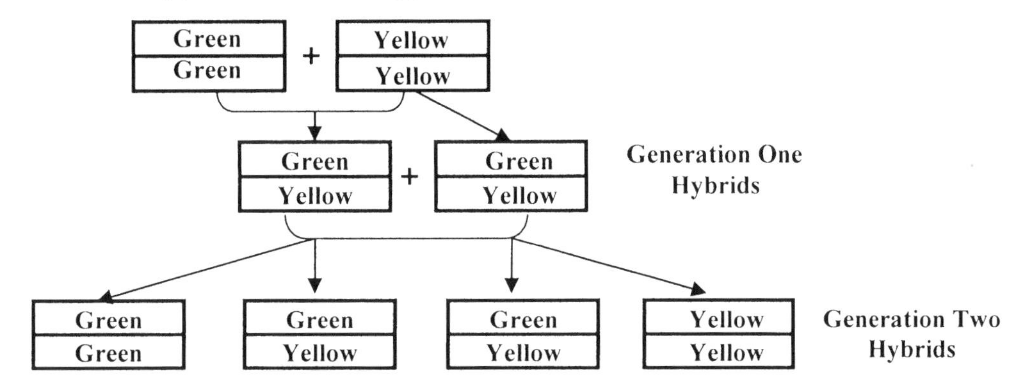

Figure 15.3 Color of Unripe Pods. Crossing a true-breed green-pod plant (green being a dominant factor) with a true-breed yellow-pod plant (yellow being a recessive factor), with each parent plant giving one factor arbitrarily, each of the next generation is a hybrid exhibiting the dominant trait, but have one of each factor. The second generation of hybrids has a 75% likelihood of having green unripe pods, only 33% of these being true-breed. There is a 25% chance of having yellow unripe pods—that is, both factors being recessive.

* Seed shape: 7324 seeds, 2.96:1 (rough and circular: wrinkled)
* Albumen coloration: 8023 seeds, 3.01:1 (yellow and orange or green)
* Color of seed coat: 929 plants, 3.15:1 (white or gray-brown, with or without violet spots)
* Shape of pods: 1181 plants, 2.95:1 (smoothly arched and not constricted or deeply constricted and wrinkled)
* Coloration of unripe pods: 580, 2.82:1 (green or bright yellow)

* Position of flowers: 858, 3.14:1 (along the stem or at the end of the stem)
* Stem length: 1064, 2.98:1(long or short)

Analysis

As the model was generated after analysis of the data in an effort to explain it, it comes as no surprise that the data are in agreement with it. There are perhaps other models consistent with the data (i.e., that could explain the disappearance of one factor and its reappearance in the second generation). Mendel did not address this issue. That his model accommodated the data that he had, to him, was sufficient for its truth. *Mendel did additional experimentation with hybrids and true-breeds as supplemental confirmation of his model* (which was ultimately found to be true). He did not, however, have a mechanism for this model of hybridization, as at his time genes were unknown.

Conclusion

At the time of Mendel's work with the evidence at his disposal, one could say that he came up with a plausible explanation for hybridization, minus a mechanism.

Monarch-Butterfly Migration

Loss of Monarch Winter Refuges

Each fall, Monarch Butterflies by the millions flock south to their overwintering grounds in Mexico. A new study indicates that someday, southward migrating Monarchs may find themselves with no place to go. The Monarchs' Mexican refuges will likely experience wetter weather within the next 50 years, and the combination of increased moisture and already cool conditions is likely to render the refuges unlivable, according to the study, led by Karen Oberhauser, an assistant professor of fisheries, wildlife and conservation biology at the University of Minnesota. The findings will be published the week of Nov. 10 in the Proceedings of the National Academy of Sciences Early Online Edition.

Monarchs currently spend winters in oyamel fir forests in mountainous areas of central Mexico. Winters there are cool, but as long as they aren't too wet, the insects survive to begin the migration north in the spring. But the knowledge that the earth is facing climate changes prompted Oberhauser and colleague A. Townsend Peterson of the University of Kansas to determine whether future winter conditions in the Mexican refuges would continue to support butterfly populations.

Using the technique of ecological niche modeling, the researchers determined the fine-scale climatic traits common to sites suitable for Monarchs. Those traits allowed them to predict current locations of Monarch populations with a high degree of accuracy. They then compared the Monarch-friendly traits to fine-scale climatic traits predicted for sites in Mexico during the next 50 years.

All current climate models predict more storms in places where Monarchs spend the winter, said Oberhauser. When the already cool Mexican forests become wetter during the overwinter-

ing period, the increased moisture will result in more frequent storms, pushing the climate outside the range of conditions necessary for Monarchs to live. "The conditions that Monarchs need to survive the winter are not predicted to exist anywhere near the present overwintering sites", said Oberhauser. "The temperatures won't change much, but the combination of coolness and increased rain will hurt".

The plight of the Monarchs is worsened by human activity in the area, she said. Logging, clearing of forests for agriculture, harvesting of wood for home fires, livestock grazing and intentional forest fires have all shrunk habitat suitable for Monarchs in the winter. Furthermore, winter storms in the Mexican fir forests in 2002 caused massive Monarch mortality, an indication that storms are an important determinant of Monarch survival.

"This study demonstrates that it is important to consider a changing climate when making conservation decisions", said Oberhauser. "Organisms need to have habitat that supports them both now and in the future. It is also an example of the far-reaching impacts of a changed climate. Organisms have adapted to survive in very specific conditions, and the combination of habitat loss and a changing climate could have large consequences".

The study represents the first use of ecological-niche modeling to predict the seasonal distributions of a migratory species under climate change. Oberhauser and Peterson will now use the technique to study how Monarchs will fare in their breeding areas within the United States. If the summer breeding areas should move north with global warming, Monarchs may need to migrate farther, Oberhauser said.

Although Monarch Butterflies may not form the cornerstone of any ecosystem, their loss will and should be felt, she said.

"If we lose monarchs, we lose a species with a unique and interesting biology, and an organism that helps many people connect to the natural world", said Oberhauser. "They're an example of how we can value components of the natural world without putting a dollar figure on them".

***Assessment of* Loss of Monarch Winter Refuges**

Model/Theoretical Hypotheses

M: The *ecological-niche model* of the oyamel fir forests of central Mexico.

H_{ENM}: Ecological-niche modeling is a viable technique that allows researchers to determine the fine-scale climatic traits that are hospitable or inhospitable for species.

Prediction

If the ecological-niche model is true, then *Monarch Butterflies may disappear entirely from oyamel fir forests in mountainous areas of central Mexico within the next 50 years.*

Background Information/Research Methods

This is the first use of ecological-niche modeling. According to this method, researchers determine the fine-scale climatic traits common to sites suitable for Monarchs, predict, after examining these traits, the current locations of Monarch Butterflies, and then map on Monarch-friendly traits to fine-scale climatic traits predicted for sites in Mexico in the next 50 years.

Ecological-niche modeling too is being tested. It is also worth noting that human activities like logging, agriculture, livestock grazing, and clearing of forest may hasten the prediction.

Data

* Winter storms in the Mexican fir forests in 2002 caused massive mortality of Monarch Butterflies. This shows storms impact Monarch survival.
* All current climate models predict more storms in the Mexican forests where Monarchs spend the winter.
* Climate models predict no hospitable conditions for the monarchs near these grounds.

Analysis

I suspect here that it is ecological-niche modeling that is being put to the test. At the very least, this forward-looking research provides some insight into whether or not this area of Mexico will have monarchs in the next 50 years or so. If the prediction turns out true, loss of monarchs will likely not pose any threat to the ecosystem.

This use of ecological-niche modeling seems a promising tool to help conservationists devise long-term strategies for keeping certain species of organisms thriving, when changes in hospitable areas can be anticipated years beforehand.

Conclusion

If indeed ecological-niche modeling is being put to the test here, then time will ultimately test just how good of a modeling technique it is.

KEY TERMS

prediction
background information
analysis
Freud's theory of dreams
Mendel's two-factor model
research methods
data
conclusion
Herschel's garden model
ecological-niche modeling

TEXT QUESTIONS

* To what extent do the three illustrations above follow the standard hypothetico-deductive approach to science? Are there instances that deviate from it?
* Using all the data that had at his disposal, what other experiments could Mendel have done to confirm his two-factor model?
* How might computer modeling be changing the practice of hypothesis-testing?

TEXT-BOX QUESTION

Compare the search for Neptune (Module 11) with the search for the planet Vulcan. How did the latter prove to be a difficulty for Newton's dynamical system and a triumph for Einstein's General Theory of Relativity?

---

1 Sigmund Freud, *Standard Edition of the Complete Psychological Works of Sigmund Freud*, Vol. 22, trans. James Strachey et al. (London: Hogarth Press, 1933).
2 Michael A. Hoskin, *William Herschel and the Construction of the Heavens* (London: Oldbourne, 1963).
3 Gregor Mendel, "Experiments with Plant Hybrids", *The Origin of Genetics*, ed. Curt Stern and Eva R. Sherwood (San Francisco: W. H. Freeman and Company, 1966).

Module 16
Competing Models

"The intention of the Holy Ghost is to teach us how one goes to heaven, not how heaven goes". Galileo, *Letter to Christina*

OFTEN IN SCIENCE several theoretical models of a phenomenon are proposed in an effort to find the correct explanation for it. Most often, these models are inconsistent with each other. When so, they are called "competing models".

Two models may be said to be COMPETING MODELS when they are given at the same time to explain some phenomenon and the explanations they give are inconsistent with each other.

This section examines two instances of competing models: one concerns contractionist and continental-drift models of Earth's crustal deformations; the other concerns geocentric and heliocentric models of Earth's position in our solar system.

Deformations of Earth's Crust

Contractionism

Following up on the work of American geologist James Hall on patterns of deformation in the Plaeozoic strata of the Appalachians, J. D. Dana (1813-1895) proposed a contractionist explanation—one that does not allow for major shifting of the earth's crust over time. Dana posited that these geological deformations were wholly explicable by the cooling of the earth, once a molten mass, over time. As it cooled, the surface began to solidify and there arose areas of crust with varying densities. High areas, which would become continents, formed a dense granitic crust, while low areas, which would become ocean beds, formed a dense basaltic crust. Continued cooling brought about contraction, where boundaried zones became stressed. Increasing stress at such zones caused downwarped troughs to form near the edges of the continents. As weathering eroded the continents, these troughs soon became filled with sediments that were folded and pushed upward as cooling and contraction continued. Thus, Dana thought the land masses that exist today, with their myriad deformities, could be completely explained by many millions of years of cooling and the forces of upheaval and erosion.

Others after Dana accepted his contractionist model, with modifications, and this was the received model of the formation and various deformations in the earth's crust well into the 1960s.

Problems with the model were noted in time. First, Dana's contractionist model posited that cooling and contraction would result in stresses that caused mostly vertical motions and vertical deformations (i.e., upwarps and downwarps). However, observation reveals that many areas of the earth's crust are being stretched, not contracted. This seems inconsistent with a model of a cooling earth where cooling results in continued contraction. Second, mountain belts reveal de-

formations that are the result of very large contractions. Dana's model allows only for small contractions through a lengthy process of slow cooling, and these chiefly in a vertical direction. Last, according to Dana's model, all mountain ranges should be roughly the same age. However, this was known even during his own day to be false.

Continental-Drift Theory

In contrast to contractionism, Alfred Wegener (1880-1930) proposed a radical, kinematic explanation of the extant land masses. Noting that the continents, from a birds-eye view, would seem to fit together if they could be pushed together (e.g., Africa fitting with North and South America), he posited that they came to be separated from what was once one giant land mass—Pangaea. About 300 million years ago, this giant continent began to break up and distinct land masses began to drift apart. However, other than the birds-eye observation of seeming fit, the hypothesis had little going for it.

It was well known during Wegener's time that fossilized plants and animals in South America matched those of the same period in Africa. Searching the paleontological literature, he discovered a paper that hypothesized that these similarities could be explained by the existence of a land bridge that connected the two continents at one time.

This land-bridge hypothesis, however, had one key difficulty. What was true of fossils common to Africa and South America was also true of fossils common to North America and Europe as well as India and Madagascar. For the majority of these fossilized organisms, travel across vast oceans would have been impossible and it seemed ad hoc to propose the existence of additional land bridges to solve the problem. Still, evidence was mounting, and continental drift now seemed not to be so absurd.

Soon more data was discovered that supported continental drift. Glacial scraping of the African and South American coastal land surfaces showed that the two continents had been close to each other during the Pennsylvanian ice-age period.

Wegener's continental-drift model also offered a different explanation for the formation of mountains than that of contractionism. For Wegener, mountains formed when the leading edge of a shifting continental mass encountered resistance. The resistance caused an upfolding of continental mass. As an example, he stated that the Himalayan Mountains were formed by a northerly drifting of India into Asia.

There was resistance to Wegener's model. For one, he never came up with a plausible mechanism for continental drift. Those mechanisms he did propose were swiftly and easily shown implausible. Thus, his model was never fully accepted by scientists of his day.

A year before Wegener's death, Arthur Holmes proposed thermal convection as the mechanism for continental drift. As the mantle of the earth heats up, it expands. This causes molten rock to shoot up to the surface to cool. This continual heating and cooling, called thermal convection, was mostly ignored at the time until work in the 1960s on ocean ridges and trenches confirmed Holmes' notion of thermal convection by demonstrating a spreading of the sea floor. This was the much needed mechanism that eluded Wegener and made plausible the continental-drift model. It is believed today that thermal convection, because of sea-floor shifting, causes the Atlantic Ocean to widen anywhere from one to 10 centimeters per year.

Analysis of Contractionism vs. Continental-Drift Theory

Models/Theoretical Hypotheses

M_C: The *contractionist model* of deformations of the earth's crust.
H_C: All geological deformations are the result of upwarps and downwarps, because of *cooling and condensing* of molten material over time.
M_{CD}: The *continental-drift model* of deformations of the earth's crust.
H_R: All geological deformations are principally the result of upfoldings and downfoldings of the earth's crust over time, cause by *resistance* to its horizontal movements.

Prediction

P_C: If the contractionist model is true, then *there will be minimal shifting of the earth's crust over time, the shifting that occurs will mostly be vertical,* and *there should not be gross deformations.*
P_{CD}: If the continental-drift model is true, then *there will be noticeable horizontal shifting of the earth's crust over time, different deformations will have different ages, there should be different degrees of deformations,* and *there must be a mechanism responsible for the crustal shifting.*

Background Information/Research Methods

It is extremely difficult to get precise measurements of crustal shifting, though the mechanisms causing such shifting have been observed and amply confirmed. The difficulty comes in the extremely slow rate of shifting.

Data

* The seeming fit of certain continents (North and South America and Africa).
* Many areas of the earth's crust are being stretched, not contracted.
* Mountain belts reveal deformations that are the result of very large contractions.
* Not all mountain ranges are the same age.
* There are common fossils in Africa and South America, North America and Europe, and India and Madagascar.
* Glacial scraping of the African and South American coastal land surfaces showed that the two continents had been close to each other during the Pennsylvanian ice-age period.
* There is a widening of the Atlantic Ocean floor of one to 10 centimeters each year.

Analysis

The data offer confirmatory evidence for the continental-drift model and give no good reason to hold on to isostatic models of the formation of the earth's crust, such as contractionism. Shifting does occur and seems to be the primary mechanism responsible for crustal deformations, though lesser mechanisms (e.g., erosion) are obviously at play. Even the small amount of shifting per year, over millions of years, would be a sufficient mechanism for crustal deformations observed today. This does not rule out some role for contractional deformations due to cooling and condensation.

Conclusion

The continental-drift model seems sufficiently justified by the data. The mechanisms by which drifting occurs are directly observable, even if the rate of shifting is hotly contested. Crustal contraction probably plays only a minor role, if it plays any, in crustal deformations.

Ptolemaic Geocentrism vs. Copernican Heliocentrism

The Aristotelian Backdrop

Copernicus' heliocentric model of the universe was published in 1543, the year of Copernicus' death. When published, it received little attention. The reasons for this lukewarm reception were many. First, it could not be proposed as a mechanically simpler model of the universe than that of Ptolemy, proposed some 1300 years prior, as both were mechanically complex and both had difficulty in explaining the observable phenomena. Next, psychologically speaking, the notion that the earth was the center of the cosmos must have been, to most, comforting. Yet it was also a notion that had the backing of Catholic exegetes. Finally, both models labored under the assumption of the truth of the principles of Aristotle's physics (see Module 3), as well as other, less basic, principles, which I lay out below.

Some Principles of the Superlunary Cosmos:
1. All celestial motions are uniform and circular.
2. All celestial objects move through the medium of aether.

Some Principles of the Sublunary Cosmos:
1. All heavy objects (earth and water, earth being heaviest) move rectilinearly toward the center of the universe.
2. All light objects (air and fire, fire being lightest) move rectilinearly away from the center of the universe (but no farther than the sphere of the moon).
3. All speed of motion is speed that is inversely proportional to the resistance of the medium through which it moves and proportional to its weight (roughly, $V \propto W/R$*).*

Ptolemaic Principles

Let us begin with the axioms of Ptolemy's geocentrism from sections one through six of his work Almagest, *Book I. For Ptolemy, as is the case with Copernicus, our solar system is the middle of the universe, which is roughly everything that can be seen with the naked eye. Ptolemy posits:*

1. The heavens move like a sphere.

2. The earth, too, taken as a whole is sensibly spherical.
3. The earth is in the center of the heavens.
4. The earth has a ratio of a point to the heavens.
5. The earth does not have any motion from place to place.
6. There are two different primary motions in the heavens:
a. There is a motion that carries everything from east to west (i.e., the motion of the "fixed stars").
b. There is a motion opposite to the first motion (i.e., west to east—the motion of the planets, the sun, and the moon).

Principle one was reasonable, as it accorded with observation. Moreover, as Aristotle said, the heavens, being divine, must move circularly. Principles two and four were not in dispute. The third and fifth principles, of course, made sense in light of Aristotle's claim that heavy things (i.e., water and earth) move to the center of the universe, where they find rest and their proper place. Last, the sixth principle was needed to account for the movements of the planets, the sun, and the moon in a direction contrary to the motion of the fixed stars.

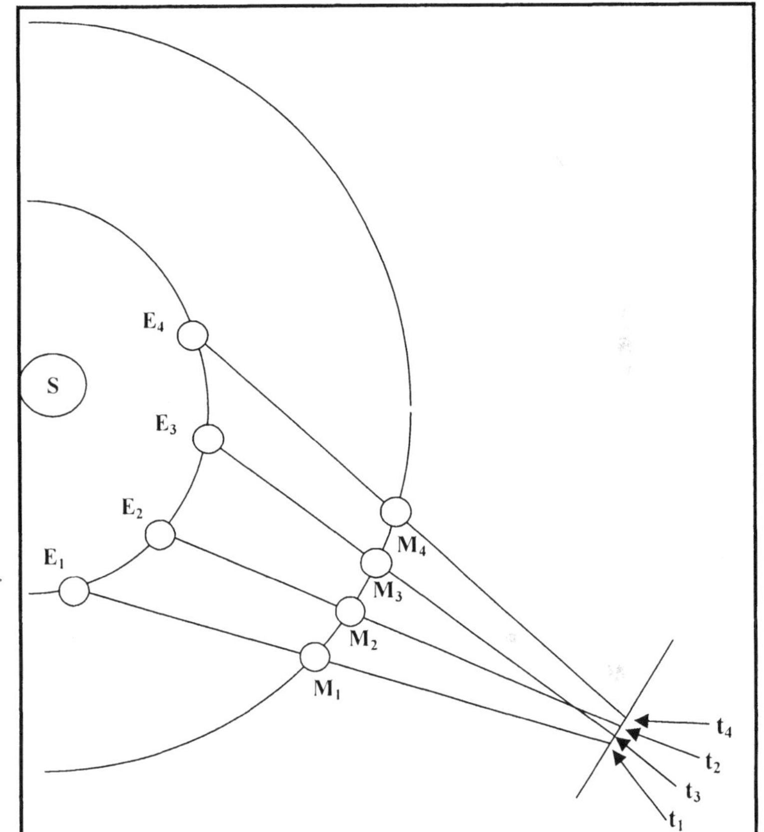

Figure 16.1 Planetary Retrogradation. This diagram shows how planetary retrogradation occurs on the heliocentric model. As Earth moves around the sun, S, outer planets like Mars, farther from the sun, orbit more slowly around their larger orbital pathway. So, the movement of Mars with respect to the "sphere of the fixed stars" sometimes seems to reverse. This retrogradation is a result of its slower motion outside of Earth and its larger orbital path and is only apparent.

Copernican Principles

Using Ptolemy's Almagest as a model, Copernicus put forth an alternative model of the universe in his De revolutionibus orbium coelestium (On the Revolutions of the Celestial Orbs). *Book I, sections one through ten, contain his fundamental principles. I list selected ones below:*

1. The universe is spherical.
2. The earth too is spherical.
3. The motion of the heavenly bodies (sun excluded) is uniform, eternal, and circular, or compounded of circular motion.

4. The earth moves circularly.
5. The heavens are immense compared to the size of the earth.
6. There are three motions of the earth.

Of these, the third, fourth, and sixth were controversial from the standpoints of Aristotelian physics and common sense. That the sun, believed to be something light, should be stationary and the earth, something thought to be heavy, should be in motion seemed absurd.

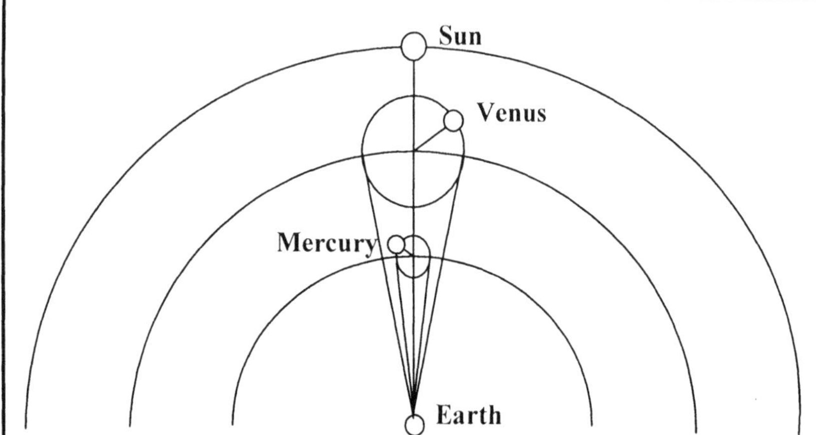

Figure 16.2. Ptolemy on Inner Planets. To explain how Venus and Mercury never appear to stray too far from the sun, Ptolemy posited that they shared the deferent arm of the sun, which (roughly) moved concentrically around the earth.

Yet not only was a moving earth deemed absurd, putting the earth in motion required that it have two additional motions! First, it had to move around its axis (roughly 1,000 m/h at the equator) to account for the perception of the daily rising and setting of the sun. Second, since the precession of the equinoxes (the slow westward motion around the ecliptic that makes a complete circuit roughly every 26,000 years) was known to Hipparchus in the second century B.C., the pole of the earth had to be given a very slow, wobbling motion.

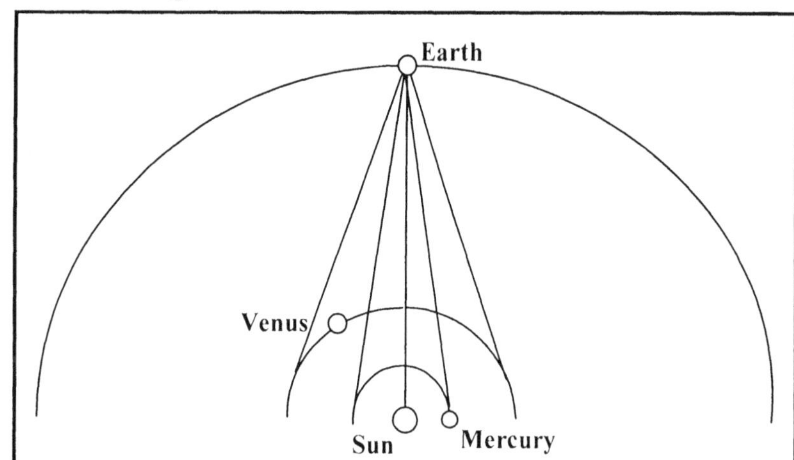

Figure 16.3. Coperniculs on Inner Planets. By positing that Earth moves around the sun and that the orbits of Mercury and Venus are inside of that of Earth, Copernicus arrives at a simple explanation of bounded elongation.

Despite these difficulties, there were certain advantages of the Copernican model. First, the observed backward path that outer planets sometimes took (i.e., planetary retrograde motion) could easily be explained by a moving earth (see Figure 16.1, p. 151). Second, heliocentrism provided a simple explanation for the bounded elongation of Mercury and Venus—the fact that both planets never stray too far from the sun. Ptolemy could only accommodate bounded elongation with much difficulty.

The mechanism Ptolemy employs (see Figure 16.2) seems ad hoc. Yet on the assumption of geocentrism, he is obliged to do just this or something like it to "save the phenomena".

Comparing Statements

Logically, propositions can be related to each other in four different ways.

1) Contradictory Claims: Two statements, α and β, are *contradictory* if whenever α is true β is false and whenever α is false β is true. For example:

α) Creation science is the true view of the origin of humans.

β) Creation science is not the true view of the origin of humans.

These two propositions are clearly contradictory in that the second is the denial of the first and, thus, the truth-value of the first will always be opposite the other.

2) Inconsistent Claims: Two statements, α and β, are *inconsistent* if both α and β cannot be true at the same time. They can, however, both be false. For example:

α) God created the first organisms on land.

β) The first organisms arose naturally out of the sea through a slow course of development.

These two propositions are inconsistent in that they both cannot be true, though they both can be false. Competing hypotheses are almost always inconsistent. That is why they are labeled "competing".

3) Consistent Claims: Two statements, α and β, are *consistent* if both α and β can be true at the same time. For example:

α) The first organisms arose from the sea through a slow course of development.

β) God is the ultimate cause of all things.

These statements seem inconsistent, but they are not. It is possible for both to be true. A creator could have set up things such that the first organisms arose from the sea.

4) Equivalent Claims: Two statements, α and β, are *equivalent* if both α and β describe the exact state of affairs—that is, whenever α is true, so too is β, and whenever β is true, so too is α. For example:

α) Zoltan is a professional athlete.

β) Zoltan makes his living playing sports.

Any conceivable world where statement one is true would be a world where statement two is true also. Conversely, any world where statement one is false would be a world where statement two is false also. The truth-value of each is always the same under all possible conditions because they mean precisely the same thing.

Heliocentrism, in contrast, offers a relatively easy explanation of bounded elongation. By switching the place of the earth and the sun (see Figure 16.3), there is no need for an auxiliary mechanism, use of the same deferent arm, to save the phenomena.

Many of the inadequacies of Aristotle's principles of physics had been known for centuries. Since the most pressing difficulties facing the Copernican model were those pertaining to inconsistencies with Aristotle's physics, it became clear that Copernican heliocentrism would only be taken seriously if Aristotelian physics could be replaced by a more accurate physics.

Galileo (see Module 10) championed this cause. He constructed a telescope for astronomical purposes, which allowed for a simple and determinate test, a crucial test, *to confirm one model and show that other is untenable.*

A CRUCIAL TEST is one that will confirm one of two competing hypotheses and disconfirm the other of the two.

A review of Figures 16.2 and 16.3 shows how this crucial test was conducted. In Ptolemy's

---

1 Brahe had come up with a third model, where the earth was the center of the universe, the sun and moon orbited the earth, and the other planets orbited the sun.

Module 17
Exercises for Scientific Models

"In theory there is no difference between theory and practice. In practice there is". Yogi Berra

EVALUATE EACH OF THE REPORTS BELOW according to the six guidelines for models below. Review the examples in Module 16 for reports containing competing models.

MODEL/THEORETICAL HYPOTHESIS: *What is the model under investigation? What theoretical hypotheses drive this model?*
PREDICTION: *What prediction(s) does the model through any of its theoretical hypotheses make?*
BACKGROUND INFORMATION/RESEARCH METHODS: *What auxiliary hypotheses here come into play? Is there relevant background information? Is there anything worth mentioning about the research methods?*
RESULTS: *What are the relevant data?*
ANALYSIS: *Are the data consistent with the prediction?*
If so, to what extent do they confirm it? Are there other models or hypotheses consistent with the data that cannot be ruled out?
If not, must the model be rejected completely or is there some non-ad-hoc way to salvage it?
CONCLUSION: *Give your verdict of the model here succinctly and precisely.*

Use outside resources, like the web, to glean further information for your research, if necessary.

A Psychologist Looks at Morality

Kohlberg's Six Stages of Moral Development

A developmental psychologist at Harvard University, Lawrence Kohlberg became famous for his work on moral development. Drawing inspiration from the philosopher John Dewey and the pioneering work of the Swiss psychologist Jean Piaget, he developed a view that moral reasoning progressed through a series of three levels and six stages (two stages per level).[1]

Level I. Preconventional Morality
Stage 1. Obedience and Punishment Orientation
Stage 2. Individualism and Exchange
Level II. Conventional Morality
Stage 3. Good Interpersonal Relationships
Stage 4. Maintaining the Social Order
Level III. Postconventional Morality
Stage 5. Social Contract and Individual Rights
Stage 6: Universal Principles

Kohlberg's initial research consisted of 72 boys, between the ages of 10 and 16, from Chicago. In later research, he expanded his research to include children, male and female, from other cities and countries.

The first level of moral reasoning, preconventional morality, is mostly found in elementary-school children. The first stage consists of moral behavior driven by an authority figure, such as a parent or teacher. Children obey such figures principally through fear of punishment. In the second stage, right behavior is deemed a matter of acting in one's own best interests.

The second level of moral reasoning is conventional morality. Its first stage involves acting in such a way to gain the approval of others. Moral situations are often two-person affairs with family or friends. The second stage involves a broadening of the conventional sphere. People obey laws, show respect for authority, and act in such a manner to preserve social order.

The third level of moral reasoning, post-conventional morality, Kohlberg claimed to observe in some, but not most, adults. In the first stage, people see themselves to be contractually bound to act in such a way to benefit the whole of a society. Though embracing different values, rational people agree that certain basic rights, like life and liberty, must be secured for all and that democratic principles work best to secure justice for all. This final stage is characterized by respect for human dignity, strictest impartiality in decision-making, and recognition that the principles of morality are universal. The influence of Kant, Rawls, and Gandhi is apparent here.

These stages were discovered by Kohlberg through observing how people reason through different moral scenarios. People progress through stages, he believed, that occur one at a time. It is not possible for someone to leap, for instance, from stage three to stage five, without passing through stage four. Critical discussion of moral reasoning therefore aims at assisting a person to see the reasonableness of the next stage of development. Following Dewey, he believed that moral discussion could promote moral progress through insight gained by cognitive conflicts at a current stage. These lessons are remembered and integrated into the next stage of development. Social interaction, then, is a key to moral conflict and conflict resolution.

One difficulty with Kohlberg's data is that it is based on what subjects say when discussing moral problems and people do not always act consistently with what they say they believe. Kohlberg acknowledged this difficulty, but he maintained sternly that his was a cognitive model that progressed through cognitive conflicts.

Another problem is that Kohlberg's stages may be culturally biased toward the Western philosophical tradition. Is such a model applicable, for instance, to non-Western cultures with different philosophical traditions? More research needs to be done here.

Fellow researcher Carol Gilligan (see textbox, below) has argued for sex-based biases in Kohlberg's work. She states that, as Kohlberg's model is derived from male-generated data, it has a decidedly male-orientation that is based on adherence to rules, fundamental rights, and assenting to abstract principles. This theoretical approach to morality excludes females, whose social interactions are less formal and more personal, real-world, and caring. Kohlberg's biases seem evident in just how women fare on his developmental scale. Though mature men typically make it to stages four or five, mature women seldom advance beyond stage three.

Finally, Kohlberg always stated that stage six was the final and highest stage of moral development, but admittedly never found enough subjects to define that category. How, then, could he be sure in the universal applicability of this stage? Might not this constitute a philosophical bias undermining his research?

Carol Gilligan's Criticism of Kohlberg

Gilligan, in 1970, began at Harvard University as a research assistant of Lawrence Kohlberg. She would soon become an ardent critic of Kohlberg.

Her criticisms are twofold. First, she maintains that Kohlberg's research is likely to be skewed in that most of his subjects were privileged, white males. She felt that this caused a biased opinion against women. Second, according to his six-stage theory of moral development, women generally fare much worse than males—thereby making the former moral inferiors to the latter.

The limits of Kohlberg's research prompted her to conduct her own research on female subjects so as to provide a more accurate picture of female moral development. Her research, often simply referred to as "difference feminism", was published in a 1982 book, *In a Different Voice: Psychological Theory and Women's Development*. According to Gilligan, men tend to think in terms of rights and justice. Theirs is a justice-orientation. Women, in contrast, tend to think in terms of care. Theirs is a responsibility-orientation. She does not argue that one tendency is superior or preferable to the other, but merely that both tendencies ought to be given equal consideration.

For Gilligan, there are three stages of moral development. First is the selfish stage where female children think the proper moral orientation is the one they in fact have—selfishness. In the second stage, guided by a consideration of care for others and by the recognition that selfishness is wrong, women turn away from selfishness and turn to a concern for others. This is conventional morality. In the final, post-conventional stage, women recognize the limits of a complete other-concern attitude and seek relationships in which they seek concern for all parties.

Gilligan's own work has undergone criticism—especially by Christina Hoff Sommer, who has challenged that Gilligan's data are problematic. In "The War against Boys", she writes:

> In September of 1998 my research assistant, Elizabeth Bowen, called Gilligan's office and asked where she could find copies of the three studies that were the basis for *In a Different Voice*. Gilligan's assistant, Tatiana Bertsch, told her that they were unavailable, and not in the public domain; because of the sensitivity of the data (especially the abortion study), the information had been kept confidential.

Summer adds that further inquiry led to a reply that the data would not be available "anytime soon". This fueled suspicion that her data were insufficient, biased, and non-replicable.

A Cosmic Alternative to Evolutionary Biology

Cosmic Evolution

In several publications in the latter part of the 20th century, Fred Hoyle and N. Chandra Wickramasinghe have maintained that the occurrence of just those conditions on earth for the development of life in Darwinian fashion is too improbable for serious scientific consideration. Instead they have proposed an alternative hypothesis: Life could have originated not on the surface of planets, like Earth, but in the large gas and dust clouds, like those of the Orion Nebula, that go into the formation of planetary systems.

Hoyle and Wickramasinghe believe that inorganic cosmic dust grains—comprising silicate, iron, and graphite—mix with organic gases to produce organic molecules that vary in their complexity. These organic molecules, then, polymerize on the surface of the cosmic dust grains. As these polymerize, the cosmic cloud increases in density and gravity and the polymerized molecules begin to collide more and more. This results in composites of grains, about the size of simple-organism cells, with chemicals needed for the formation of life. From here, natural selec-

tion takes over, as it does with organisms on Earth, and the simplest, self-replicable composites of grains, bound by organic polymer coatings, begin to dominate. When the conditions are right, cell membranes form—each with the essential ingredients of life. Finally, again under the right conditions, these "first cells" are then conveyed via meteorites and comets to planets like Earth, where they act as the seedlings of life.

Critics have pinpointed difficulties with the model. First, to date there is no record of any cosmic debris with prebiotic materials on Earth. Second, the sun's ultraviolet radiation, cosmic radiation, and particle zones—like that of the Van Allen Belts—would destroy all such prebiotic materials as they traveled through outer space. Last, there is considerable reason to question whether Hoyle and Wickramasinghe have offered a solution to the problem of the unlikelihood of prebiotic materials forming on Earth at all. After all, is their model any more likely than others they are criticizing?

The Laws of Thermodynamics

Thermodynamics is the study of the laws that govern conversion of energy, the availability of energy to do work, and the direction of the flow of heat. There are four laws of thermodynamics, two of which are well-known and given below.

1ST LAW: In an isolated system anywhere in the universe, the total energy, both potential and kinetic, cannot be changed unless the system ceases to be isolated.

2ND LAW: Heat cannot be transferred from one body to a second at a higher temperature without producing some other effect, or the entropy (i.e., disorder) of a closed system increases with time.

The Phlogiston Model of Combustion

Priestly's Phlogiston Model

When Joseph Priestly (1743-1804) analyzed the gaseous products of the decomposition of metal nitrates and mercury oxide, he developed the view that the gas produced was "dephlogisticated air". A new model of combustion was formed. According to Priestly, phlogiston is a mysterious, almost weightless substance that is removed from combustible substances, like metals, during combustion. As metal is calcined, it turns to a powdery substance called a calx (i.e., an oxide). The slight loss of weight is explicable as the loss of phlogiston in the air. When ore is smelted, however, the process reversed. As charcoal, thought to contain an abundance of phlogiston, is burned with this powdery calx, phlogiston passes from the charcoal to the calx and restores the metal. The slight residue that charcoal leaves behind is explicable by its being composed of much phlogiston. According to Priestly, fires in enclosed spaces burn out because the air becomes saturated with phlogiston.

Chemists noticed however that the process of combustion for some metals resulted in a calx that was heavier than the initial metal. That, of course, should not have happened. Adherents of the phlogiston model, of which there were many different versions, tried to meet this problem by positing that phlogiston had negative weight in some metals. Thus, when the phlogiston was released into the air, the metal would become heavier. Another difficulty was that mercury could be returned to a metal by heating alone. It did not require a phlogiston-rich source like charcoal.

To overcome such difficulties, Antoine Lavoisier (1743-1794) proposed an alternative hypothesis: Combustion is a chemical reaction that requires air—specifically oxygen. Thus, the increase in weight in some metals is explained not by phlogiston with negative weight, but by those metals combining with air. Enclosed fires burn out not because the air is saturated with phlogiston, but because of lack of air. In his work on chemical reactions, Lavoisier also proposed the conservation of mass—that the mass before a reaction must be equal to the mass after a reaction. This helped to make chemistry a quantitative science.

Wired for War?[2]

Killer Chimps and the Beginnings of War

Researchers, studying chimpanzees' behavior, have documented numerous incidents of chimpanzees ganging up on, hunting down, and killing other chimpanzees. Some scientists maintain that such behavior is hard-wired in the animals and that studying it could help us understand the biological roots of our own savageness. Others argue that the incidents are not biologically, but socially, rooted. Certain elements of modern civilization, they maintain, are responsible for the violence.

Two new reports of violence, documenting 11 killings and additional maimings in two chimpanzee communities totaling over 200 members, have recently surface to reinvigorate the debate. "Lethal coalitionary aggression is part of the natural behavioral repertoire of chimpanzees", writes David Watts of Yale University in New Haven, Connecticut, in one such report, at the Annual Meeting of the American Association of Physical Anthropologists. The other appeared in the International Journal of Primatology.

Chimps live in "communities". Reports of chimp violence occur mostly when chimps wander near or into the territory of a neighboring chimp community. If a group of chimps come across a chimp from another community, who is alone, they sometimes attack.

One hypothesis—the balance of power hypothesis, is that war has evolutionary roots. According to this view, the mutual group violence of animals enables them to win resources and territory, which in turn leads to greater chances of survival and passing on their genes. In short, evolution favors animals who kill and get away with it. "This makes grisly sense in terms of natural selection". said Richard Wrangham, a professor of anthropology at Harvard University in Cambridge, Mass., and the author of the hypothesis.

Human and chimp wars differ in major ways, Wrangham says. Chimps seem to be better judges of what they can get away with. While chimpanzees' battles are generally quickly resolved, human wars drag on for many years.

There are similarities. Wrangham adds, "If we as human males feel we are in a position to kill safely, then we're easily induced to do it. The old principle of attacking safely is still there". Among hunter-gatherers, "the surprise raid is the typical pattern. The aim is to get together a small group of men who go off, and find a helpless victim, kill them, and run away again".

Other scientists disagree with Wrangham. Brian Ferguson, a professor of sociology and anthropology at Rutgers University says chimp violence has been exaggerated. Reports of chimp violence came from Tanzania in the 1970s, when whole chimp communities were said to be eliminated by others. He adds, "In many cases, all we know for sure is that some chimps disap-

peared", and some scientists have "a tendency to take a disappeared chimpanzee as a killed chimpanzee". Ferguson adds that, human interference, like logging, may be a factor. Such activities limit those of chimps by restricting land and food resources and forcing them to fight for the dwindling remains. For instance, the forest cover has vanished all around Kibale National Park, home of the chimps in Watts' study, who accounted for eight of the 11 killings mentioned in the two new reports. "They're totally hemmed in now", Ferguson said. "It's a very human kind of situation: a population that's growing, that can't go anywhere, may be beginning to run down its resources".

If Ferguson is right, war is not biologically rooted, but a product of modern civilization—a view popularized by Jean-Jacques Rousseau in the 18th century. There is some evidence for the Rouseauian view. Researchers who examined over 5,000 Native American skeletons found that those skeletons, dating after the landing of Christopher Columbus, showed a 50-percent higher rate of traumatic injuries than prior ones. They believe the violence in the bones may be the result of diminution of their territories, due to White people's expansion.

Wrangham disputes those findings that claim to support the Rousseauian view. He argues that Kibale Park covers more than 700 square kilometers and the chimps in it multiply at a rapid or slightly more than average rate. This shows that their resources are plentiful. "They actually seem to be very well off in terms of their food supply". Moreover, he dismisses Ferguson's idea that researchers are counting too many unconfirmed disappearances as killings. Two-thirds of the 49 killings documented to date were either directly seen or inferred from clear evidence such as chimps prancing around a brutalized corpse. Only the remaining 16 are classified as suspicious disappearances.

Other scientists dispute the balance-of-power hypothesis, because mutual killing among animals besides chimps and humans is rare. None has been found among bonobos—apes more closely related to chimps than humans are. Wrangham says that may be because only particular social structures, like a mixture of social communities with small and often-changing subgroups, make killing easy. He states, "It is a finely tuned strategy, used on occasion when killers are able to kill at very low risk to themselves".

Dolphin Games[3]

Serious Play for Evolutionary Success?

Researchers Stan Kuczaj and Lauren Highfill observed some intriguing games played by rough-toothed dolphins, while they were snorkeling off the coast of Honduras. Two adults and a youngster were passing a plastic bag back and forth. When the adults passed it to the youth, they placed it just in front of the youth's mouth, as if to make it easier to catch. Their observations were published in the journal, Behavioral and Brain Sciences.

Kuczaj and colleagues studied a group of 16 captive bottlenose dolphins for five years and also observed wild dolphins. Their research shows that dolphin games exhibit remarkable cooperation and creativity. They deliberately seem to make their games difficult, and difficult games may be a key in the development of culture and evolution in many animals. A psychologist with the University of Southern Mississippi, Kuczaj writes that games "may help young animals learn their place in the social dynamics of the group". He adds, "The innovations produced during the interactions of young animals may be important sources for the evolution of animal traditions, as well as the adaptations that may lead to more successful individuals and species".

Evolutionary biology maintains that new species emerge slowly through chance variations that are beneficial for organisms. These organisms, having a better chance of surviving and passing on their genes, pass on those variations to the population. Over time, the result is a new species. Thus, it may be that animals that may have inherited a predisposition to play, because "it helps animals gain knowledge of the properties of objects, perfect motor skills, and recognize and manipulate characteristics of [their] environment". Play—especially difficult play—must be biologically important, because many animals, especially dolphins, play at the risk of loss of life and limb, and dolphins seem to make their play very difficult.

Kuczaj's research showed that the captive dolphins "produced 317 distinct forms of play behavior during the five years that they were observed". One calf became adept at "blowing bubbles while swimming upside-down near the bottom of the pool and then chasing and biting each bubble before it reached the surface". Kuczaj adds, "She then began to release bubbles while swimming closer and closer to the surface, eventually being so close that she could not catch a single bubble. During all of this, the number of bubbles released was varied, the end result being that the dolphin learned to produce different numbers of bubbles from different depths, the apparent goal being to catch the last bubble right before it reached the surface of the water. She also modified her swimming style while releasing bubbles, one variation involving a fast spin-swim. This made it more difficult for her to catch all of the bubbles she released, but she persisted in this behavior until she was able to almost all of the bubbles she released. Curiously, the dolphin never released three or fewer bubbles, a number which she was able to catch and bite following the spin-swim release". She may have been blowing more bubbles than she could easily catch and bite to complicate her play. "These observations are consistent with the notion that play facilitates the development and maintenance of flexible problem solving skills. If this is true, play may have evolved to enhance the ability to adapt to novel situations".

Dolphins of all ages played games, Kuczaj says, but most new games were made up by youngsters. That suggests a sort of "dolphin culture". That other animals may have a culture is a relatively new idea, founded on observed strategies for using tools by chimpanzees and other primates that get passed on to offspring. Now dolphin culture is now an object of scientific investigation. "The ability to invent novel play behaviors and the ability to learn from the behaviors of others may be related to the creation and maintenance of animal traditions and ultimately to the survival of species".

Jane Goodall (1934-)

Ethologist Jane Goodall is perhaps the world's leading authority on chimpanzees. Her 45 years of research taught us much about the intelligence of wild chimpanzees—their behavior, learning, thinking, and culture. One of her most significant discoveries is that chimpanzees, like humans, use tools to achieve ends—like poking blades of grass into termite mounds. Prior to this, it was believed that only humans used tools. In fact, tool-use was thought to be the defining characteristic of human beings. Goodall also discovered that chimpanzees hunted red colobus monkeys. She adopted the convention of naming the animals she studied, instead of assigning each a number. Assigning numbers was thought to be an important part of detaching oneself from one's research.

Goodall has won numerous awards for her research and humanitarianism in her career—for instance, the Ghandi/King Award for Nonviolence and the Kyoto Prize. She heads the Jane Goodall Institute for research on and education and conservation of wildlife.

Evaluation of **Kohlberg's Six Stages of Moral Development**

Model/Theoretical Hypothesis

Prediction

Background Information/Research Methods

Results

Analysis

Conclusion

***Evaluation of* Cosmic Evolution**

Model/Theoretical Hypothesis

Prediction

Background Information/Research Methods

Results

Analysis

Conclusion

Evaluation of Priestly's Phlogiston Model

Model/Theoretical Hypotheses

Prediction

Background Information/Research Methods

Results

Analysis

Conclusion

Evaluation of Killer Chimps and the Beginnings of War

Model/Theoretical Hypotheses

Prediction

Background Information/Research Methods

Results

Analysis

Conclusion

Evaluation of Serious Play for Evolutionary Success?

Model/Theoretical Hypotheses

Prediction

Background Information/Research Methods

Results

Analysis

Conclusion

KEY TERMS

Kohlberg's six stages of morality
Phlogiston Model of combustion
Rousseau on war
laws of thermodynamics
conservation of mass

TEXT-BOX QUESTION

What does Gilligan's criticism of Kohlberg and Sommer's criticism of Gilligan's work tell you about the politics of science?

---

1 Lawrence Kohlberg, *Essays on Moral Development: The Psychology of Moral Development*, Vols. I & II (San Fracncisto: Harper and Row, 1981.
2 From *World Science*, February, 2005.
3 *World Science*, Nov. 9, 2005.

SECTION SIX

Statistical Hypotheses

Module 18
Estimation & Proportion

"If a man has one hundred dollars and you leave him with two dollars, that's subtraction". Mae West

PRIOR TO THIS SECTION, we examined theoretical models and focussed on hypotheses that, as parts of theoretical models, were universal in scope.

All dreams are wish fulfilments.
Sleep has a restorative function.
The solar system is Copernican heliocentric.

The underlying assumption was that, if the universe was a deterministic system (or nearly so), universal hypotheses and theoretical models could at least partly explain that system.

Yet it is obvious that in the practice of science, scientists often employ other types of hypotheses: causal and statistical claims.

Causal: Smoking causes lung cancer.
Statistical: People who regularly attend church have a 25% reduction in mortality.

Statistical and causal hypotheses are the driving forces behind controlled laboratory experiments. As we shall see when we move to the next section, causal hypotheses are special sorts of claims that are, in some degree, philosophically troublesome, but practically indispensable. In the modules of this section, however, the focus is on statistical hypotheses.

Statistical Hypotheses Based on Proportion

Suppose Nathan D. Wizdum, professor of philosophy at Culdesac College, wishes to know one day just how many of his students are avid readers ("avid" understood here as "reading five or more books per year other than those required for school"). The easiest way to find out would be to ask all students, on a day where everyone is in attendance, whether or not they read five or more books each year. The answer, then, might be a simple proportion: 21 of 35 students (60%), say, meet this criterion. Yet what if he wanted to know how many of the students attending Culdesac College, a school of 5,000 students, are avid readers? He, of course, could pass out a questionnaire to each professor in each class and tally the data. This, however, would be very time-consuming and difficult to manage, and there would be no guarantee that the data would be complete. A simpler way would be to take a random sample of the students, some proper subset of the actual population of students currently in attendance, and question them. This would certainly entail some margin of error, but with a large enough sample size and a randomly selected sample, he could be assured that the results he obtains are representative of the sample.

Suppose Professor Wizdum randomly selects 500 students and finds out that exactly 333

A VARIABLE is a specific property that each member of a particular population under study has.

A VALUE is the way in which the member of a particular population under study exhibits a specific property.

A PROPORTION is a simple statistical relationship in which only two values (e.g., not avid vs. avid) of one variable (e.g., reading propensity) are being considered.

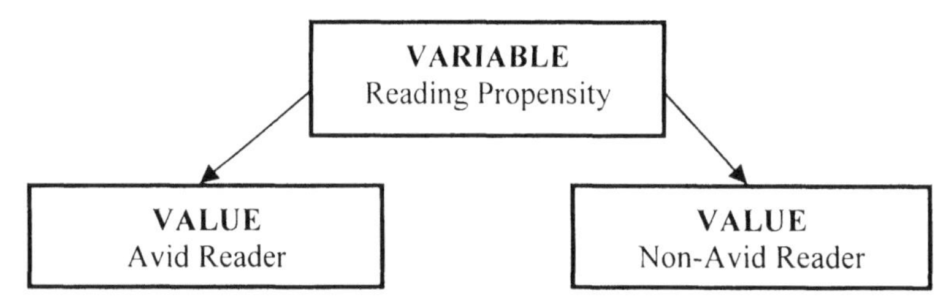

Figure 18.2 Example of Variable and Two Values

As, Professor Wizdum was studying reading propensity—a specific variable that could in principle take on different values—we have only two: avid readers and not avid readers (Figure 18.2). (Obviously, there could be more. We could, for instance, break "avid readers" into "average reader" and "light readers", if we so choose.) "Note here that these values are exclusive and exhaustive. As he studies only one variable that has exactly two values, the statistical relationship he analyzes is one of mere proportion. He asks: What percentage of all current students are avid readers?

To evaluate models based on proportions (or any other type of statistical model), we first need the appropriate tools. Here we turn to analysis of data. This takes us to the probability calculus.

Assessments based on probability allow us to form reasonable, though not infallible, judgments about populations based on samples that would not be readily available otherwise. To illustrate, let us return to Professor Wizdum's sample of 500 students at Culdesac College. Of the 500 samples students, we know that 333 (67%) are avid readers. Yet what we really want to know is the actual number of avid readers at the college. So, we ask, "What percentage of the population actually has that value (avid reader) of the variable (reading propensity) under investigation?" To answer this, we first must say something about the science of *estimation*.

Estimation

Given a random selection of the sample from the population, even if we have no information about the size of the population, the size of the sample gives us a measure of the likelihood that the results found in the population will be transferable to and representative of the population itself. The results here, as we are dealing with probability, are always fallible; there is always the possibility of error. Still a random sample and the size of this sample will allow us to speak with some confidence about how well this sample represents the intended population. To do so, we shall have to say something about a sampling distribution first.

Suppose a friend has a black vase filled with ancient coins, each being roughly the same size and weight, and exactly half of them have a likeness of Alexander of Macedon on them, while the other half have depictions of various other figures from Hellenic antiquity.

Figure 18.3. Alexander Coin. Ancient tetradrachm (16.4 grams) in honor of Quaestor Aesillas of Thessalonika (c. 90-75 B.C.). Here Alexander of Macedon, with horn of Ammon and flowing hair, is depicted on the front.

Suppose now you select 10 coins from the vase, each time replacing the coin you get and then mixing up the coins. On each try, the probability of getting a coin with the face of Alexander on it is 0.5, or pr(A) = ½. We get the following range of possible distributions for 10 trials (Figure 18.5), where A_0 represents 10 selections, each of which fails to be a coin with Alexander on it, A_1 represents the number of ways in which one could get *exactly one* coin with Alexander on it in 10 trials (A on the first pick or on the second or on the third, etc.), A_2 represents the number of ways in which one could get *exactly two* coins with Alexander on it in 10 trials, A3 represents the number of ways in which one could get *exactly three* coins with Alexander on it in 10 trials, and so on.

$pr(A_0)$ = 1/1024 (.001)
$pr(A_1)$ = 10/1024 (.01)
$pr(A_2)$ = 45/1024 (.04)
$pr(A_3)$ = 120/1024 .117)

$pr(A_4)$ = 210/1024 (.21)
$pr(A_5)$ = 256/1024 (.25)
$pr(A_6)$ = 210/1024 (.21)

$pr(A_7)$ = 120/1024 (.117)
$pr(A_8)$ = 45/1024 (.04)
$pr(A_9)$ = 10/1024 (.01)
$pr(A_{10})$ = 1/1024 (.001)

Figure 18.4. Distribution: 10 Trials. Distribution of all possible outcomes in 10 selections, with replacement.

The data may be graphed below as a sampling distribution.

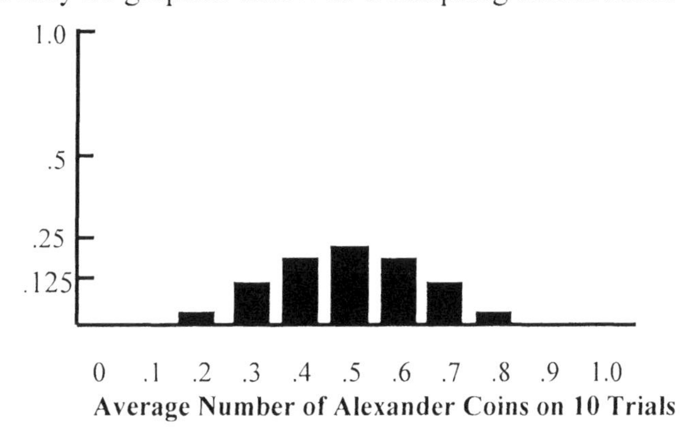

Figure 18.5 Graph of distribution of all possible outcomes in 10 selections, with replacement, where pr(A) = ½.

If we set up a distribution for 100 trials with replacement (Figure 18.6, below), we get a distribution with a similar shape, though the relative frequency of each diminishes significantly and the distribution of most probable outcomes clusters much more toward the middle—0.5.

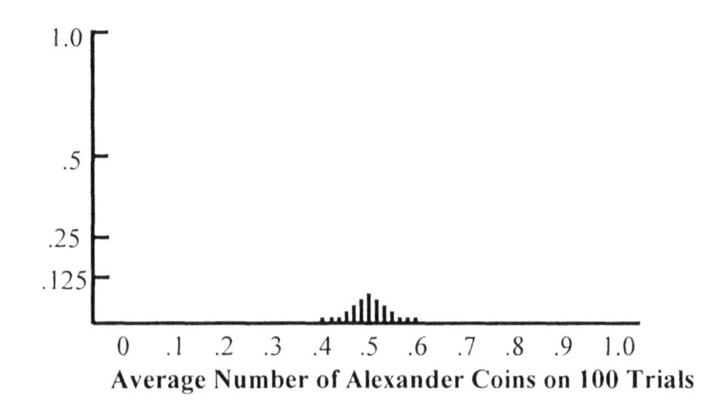

Figure 18.6 Graph of distribution of all possible outcomes in 100 trials, with replacement, where pr(A) = ½.

A pattern is apparent. The greater the number of trials, the greater is the tendency for the distribution to cluster around the actual probability of the population—here, known to us in advance. In each distribution, the greatest possible frequency—here 0.5—is called the *expected frequency*.

We see, in general, that the larger the number of trials, the less probable the expected frequency will occur in a given sample. In the distribution of 10 trials, the expected frequency is 0.5 (or five of 10 Alexander coins) and the likelihood of this actually occurring in 10 picks is 0.25. In the distribution of 100 trials, the expected frequency is still 0.5 (50 of 100 Alexander coins), yet the likelihood of this actually occurring is 0.08—significantly less than 0.25. Nonetheless—and this is the point of having a large sample—with a large sample, one can be relatively sure that the *observed frequency* of an actual sample, what we actually get when we pick 10 coins, will be *very close* to the expected frequency, whereas the large spread of a distribution in a small sample gives no such certainty.

If we investigate some value in that population, say avid reading propensity, it is clear the population has at any one time an exact number of subjects that have this value, which we shall designate by "F_P" (i.e., actual frequency of the value in a population). Yet as populations are often impossible to work with, so we need to work with samples from that population. Each sample, as a proper subset of its population, has its own frequency, which we shall designate by "F_S" (i.e., observed frequency of the value in the sample). In most cases, especially in large samples, it is very unlikely "F_S" will be *exactly the same as* "F_P". Still a large and varied sample can give us certainty that "F_S" will be *very close to and, thus, highly representative of* "F_P". We might not be able to say that "F_S" is equivalent to "F_P", but in many cases we will be able to say "F_S" is very nearly equivalent to "F_P". In other words, we are willing to sacrifice the likelihood of "F_S" being equivalent to "F_P", which is improbable in most samples, for "F_S" being very near "F_P", which is very probable in large and varied samples. The next question is this: Just what level of certainty do we need?

This is where the notion of *standard deviation* from the expected frequency comes into play. The distribution we have observed above is called a "bell" curve in that it is shaped like a bell and the distribution is symmetrical on each side of the expected frequency. Distributions where the expected frequency is not 0.5 differ slightly from this ideal bell shape, but not so much as to cause us concern when dealing with deviations from the expected frequency. Now statisticians have adopted by convention a definition of one standard deviation to contain 67% of all sample frequencies (using the expected frequency as the mean), while two standard deviations contain

95% of the possible sample frequencies around the expected frequency (Figure 18.7).

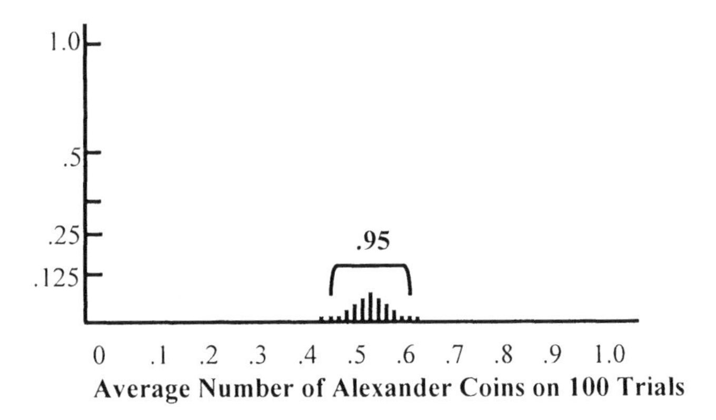

Figure 18.7. Two Standard Deviations for 100-Selection Distribution. Graph of distribution of 100 trials, with replacement, where pr(A) = ½. Here two standard deviations range roughly from 0.4 to 0.6. With more trials, the distribution clusters more toward the expected frequency.

Just how does this relate to samples and populations? Let us return to Professor Wisdum's sample of 500 students, 167 (F_S = 33%) of which were avid readers. To what extent is F_S here representative of the population? Taking 0.33 as the expected frequency or the mean of a sampling distribution of 500 trials, we can figure out with a confidence factor of 95% that this value for F_S is within two standard deviations of actual value F_P in the population. What that means that if we should take 95% of all sample frequencies around F_S (a distribution around an actual sample), we would have a 95% assurance or a *confidence interval* of 0.95 that F_P is within these parameters.

Here is a chart that we shall use throughout this section on statistical models as well as the next section on causal models.

N_S	ME	Ns	ME
10	+/-.32	1,000	+/-.03
20	+/-.22	2,500	+/-.02
30	+/-.18	5,000	+/-.01
40	+/-.16	10,000	+/-.01
50	+/-.14	25,000	+/-.006
60	+/-.13	50,000	+/-.004
70	+/-.12	100,000	+/-.003
80	+/-.11	200,000	+/-.002
90	+/-.11	300,000	+/-.002
100	+/-.10	400,000	+/-.002
250	+/-.06	500,000	+/-.001
500	+/-.04	1,000,000	+/-.001

Figure 18.8. Margin of Error for Sample Sizes. Estimated margin of error for selected sample sizes, based on pr(F) = ½. This chart is roughly applicable as well to samples where pr(F) is other than ½.

If we apply this chart to the sample of Professor Wizdum, we form the following assessment of his sample (Figure 18.9). Given that $N_S = 500$, a relatively large sample, we have a ME of +/-.04.

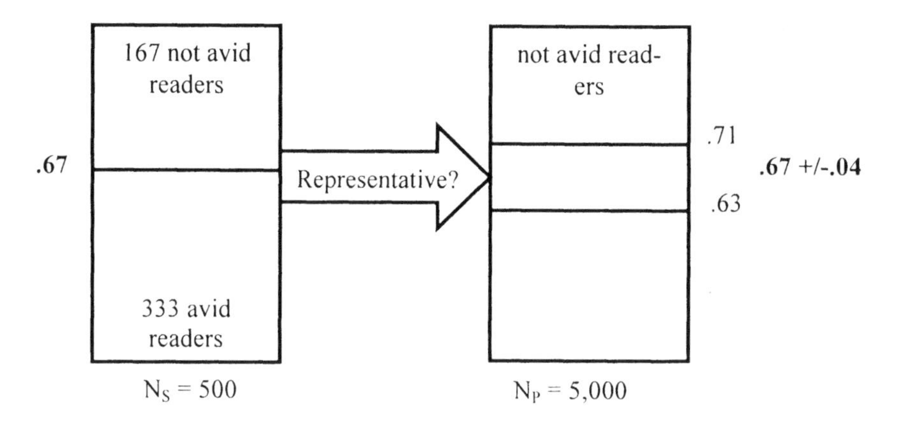

Figure 18.9 Culdesac Students: Sample and Population with ME

What this means is that there is a 95% chance that the actual population will be some percentage between 63% and 71%, inclusive. In other words, there is only a 5% chance that the number of avid readers at Culdesac College is between 0-62% or 72-100%. Given this, we may reformulate our statistical generalization more precisely thus:

> 67% of the 500 sampled students at Culdesac College are avid readers.
> So, 67% (+/- .04) of all students at Culdesac College are avid readers.

In most scientific studies, this 95% confidence interval is all that is required for reliability. In more stringent experiments, there may be a requirement that the data from the sample are accurate within three standard deviations, but this need not concern us here.

Formula for Margin of Error

One can use this general formula to calculate the margin of error for *one* standard deviation:

$$\left(\frac{(F)(1-F)}{n}\right)^{1/2}$$

Here "F" represents the frequency (use .5) and "n" represents the sample size. To get two standard deviations, as is traditionally used, multiply the end result by two.

Key Terms

population
proportion
value
estimation
observed frequency
sample
variable
Quantum theory
expected frequency
standard deviation

margin of error confidence interval

Text Questions

* What are the benefits of using samples to glean information about populations? What are some of the possible drawbacks?
* Why is it that a larger sample is more representative of a population but less likely to be exactly representative of it? Is this a benefit or weakness of large samples?

Text-Box Question

How has the notion of the discontinuity of energy created difficulties for a common-sense view of subatomic realities? Why have scientists introduced the word "quantum" to reflect these difficulties?

Module 19
Distribution & Correlation

"The statistics on sanity are that one out of every four Americans is suffering from some form of mental illness. Think of your three best friends. If they're okay, then it's you". Rita Mae Brown

THE PREVIOUS MODULE EXAMINED hypotheses based on proportions, with one variable and two values, through statistical tools given by the science of estimation. In this module, we look at hypotheses based on distributions, with one variable and more than two values, and those based on correlations, where there are two variables, each of which has two values.

Statistical Hypotheses Based on Distribution

Hypotheses concerning distributions are slightly more complex than those of proportions. Where hypotheses based on proportions have one variable and two values, those of distribution have one variable and more than two values.

A DISTRIBUTION is a statistical relationship in which more than two values of one variable are being considered.

A large truckload of limes comes to a Delmer Haines Wentworth's Fruit Market. The manager, "Slim" Smithers, is willing to purchase the truckload of limes at the price that the driver proposes, so long as no more than 10 percent of the limes are rotten. Working on the assumption that the limes were placed into their wooden crates in no particular manner, he randomly picks out one crate to get an assessment of the quality of the limes. There are 100 limes in this crate. Of these 100 limes, 61 are ripe, 20 are rotten, and 19 are unripe. Based on what we know of this sample, should Slim purchase the truckload of limes? (Figure 19.1 below).

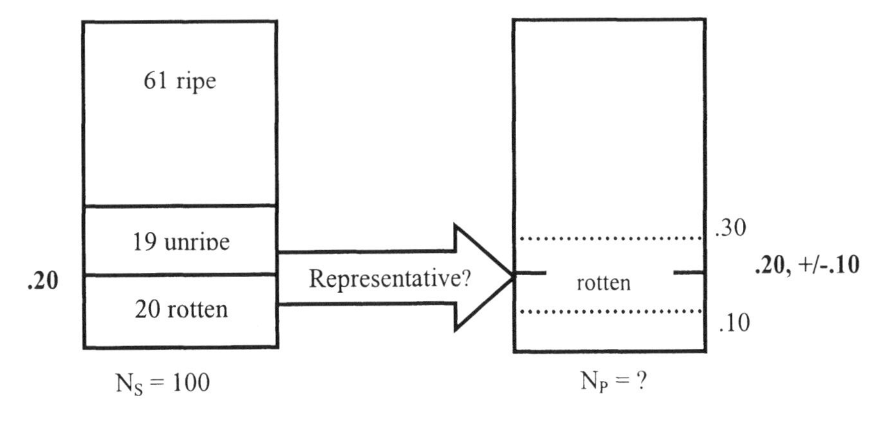

Figure 19.1 Sample of 100 Limes with ME. Given an ME of +/-.10, we can say that anywhere from 10% to 30% (dotted lines) of the limes will be rotten.

In this example, we focused only on one value, rottenness, of the variable, state of the limes. We could analyze each of the other values, unripeness and ripeness, should we so choose.

Statistical Hypotheses Based on Correlation

Correlational data are more cumbersome than proportional and distributional data, but of considerable value in scientific investigation. For instance, all cases of suspected causal relationships begin with correlational data.

> **A CORRELATION is a relationship between two variables such that a change in one variable is accompanied by a change in the other.**

A correlation can be *positive*, when an increase/decrease in the frequency of one variable results in an increase/decrease in the frequency of the other, or *negative*, when an increase/decrease in the frequency of one variable results in a decrease/increase in the frequency of the other. When two variables are not related, they are said to be *uncorrelated*. As we shall see below, rarely shall we be in position to say categorically that two variables are uncorrelated. In troublesome cases, the best we can do is state our uncertainty. In all cases, however, the data themselves will point to the degree of scepticism we should embrace in drawing out the conclusion.

Let us now take a closer look at these three different patterns of data: a positive correlation, a

Election 2000: The Florida Debacle

The 2000 presidential election between Democrat Al Gore and Republican George W. Bush was the closest in U. S. history. George W. Bush proved the eventual winner with 271 electoral votes and 50,456.169 popular votes. Al Gore, though he won the popular vote with 50,996,116 popular votes, lost the election with 266 electoral votes. Closeness notwithstanding, this was also the most controversial election. Why the controversy? The state of Florida, whose 25 electoral votes would decide the presidency.

The problems with Florida were numerous. I list just a few. First, on the night of the election, television news media called Florida for Al Gore around 9:00 pm EST, while voters in the western panhandle (which is in the Central Time Zone) of the state were still voting. This may have depressed the voter turnout in that region, which is mostly Republican. In addition, there were roughly 179,855 ballots that were not officially counted, because of difficulties with voting machines in certain counties with high African-American and Hispanic populations. Moroever, 57,700 votes were not counted, as these voters were listed as felons on a "*scrub list*", though many were not felons at all. Last, a substantial number of oversees ballots were either missing postmarks or filled out so that they were invalid under Florida law. Democrats, noticing that these were mostly military ballots, argued that these should not be counted.

After numerous court challenges and recounts, the U.S. Supreme Court ultimately ruled that the recount procedure was unconstitutional, since it was not to be conducted statewide. Though unsatisfied with the ruling, Gore ultimately conceded the victory to Bush.

These embarrasments have led to a considerable push toward election reforms—the most critical of which may center on the infamous "butterfly ballot", which was in use in Palm Beach County. The difficulty with the ballot was that candidates, who wanted to vote for Gore, may have accidentally voted instead for Reform Party candidate Pat Buchanan. Democrats were listed second in the left column, while a hole in the second circle was a vote for Buchanan. These ballots were discarded—5,330 votes of which were for Gore and Buchanan, while 1,631 were for Bush and Buchanan (*Palm Beach Post*). Other problems aside, this cost Gore the presidency.

With Florida decided by the slimmest of margins, the presidency went to Bush. Two things seem clear: Individuals, though voting, can make a huge difference in global events and samples are not infallible.

negative correlation, and cases where it is impossible to tell whether the values are correlated or not.

Positive Correlation

Let us assume a random survey of both 500 American males and females on cigar smoking, where 80 of the 500 (.16) males reply that they smoke cigars and five of the 500 (.01) females reply similarly. With a ME of +/-.04 for each, we show how the variables are related below.

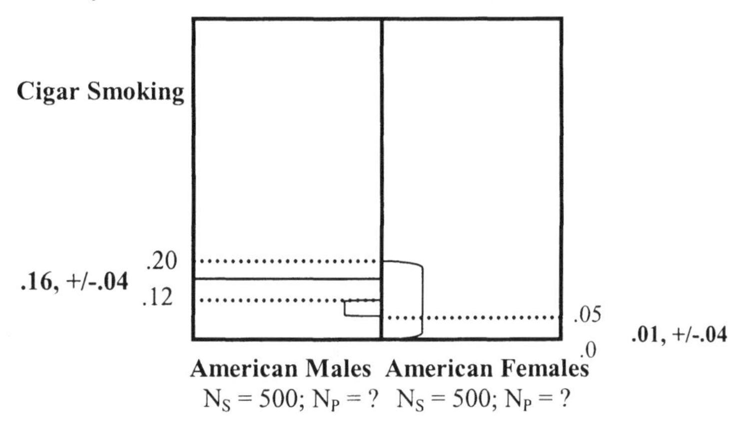

Figure 19.2 American Cigar Smokes: Males vs. Females

As there is no overlap between the range of probable values for the actual percentages of American male and female cigar smokers and the sample size is fairly large, we can state firmly that, among Americans, there's a positive correlation between being male and smoking cigars. Subtracting the lowest parameter in the 95% confidence interval of female smokers (.0) from the highest parameter of male smokers (.20), we get 20% or 0.20. Subtracting the highest parameter in the 95% confidence interval of female smokers (.0) from the lowest parameter of male smokers (.20), we get 7% or 0.07. So these data only suggest a positive correlation and the strength of the positive correlation, *being an American male is positively correlated with smoking cigars*, is likely somewhere between .07 and .20 (1.00 being a perfect or one-to-one positive correlation). In other words, a relationship exists, but it is not strong.

And so:

> **A POSITIVE CORRELATION is a relationship between two variables such that a increase in one variable is accompanied by an increase in the other and a decrease in one variable is accompanied by a decrease in the other.**

Negative Correlation

Suppose next that savvy Sebastian "Big Time" Terlick wishes to try a risky new promotional scheme to promote an upcoming art exhibit at the Detroit Institute of Arts. He wishes to promote the art exhibition at the monster-truck show at the Pontiac Silverdome, where 50,000 people are expected to attend. From a sample of 100 who admit to being monster-truck mavens, he finds that 2% express genuine interest in art. From a sample of 300 who state they are not monster-truck mavens, he finds that that 73 (.24) express a genuine interest in art.

Here again we find no overlap between the two intervals. Thus, we have a correlation. Yet the correlation is not positive, but negative. Subtracting the lowest parameter in the 95% confidence interval of those who are not monster-truck mavens (.18) from the highest parameter of those who are monster-truck mavens (.12), we get a measure of the least possible strength of the correlation, -6% or -0.06. Subtracting the highest parameter in the 95% confidence interval of those who are not monster-truck mavens (.30) from the lowest parameter of monster-truck mavens (.0%), we get -30% or -0.30. We can conclude that *being a monster-truck maven is negatively (or inversely) correlated with having a genuine interest in the arts* and that the strength of the correlation is somewhere between -.06 and -.30 (-100% being a perfect or one-to-one negative correlation). Therefore, it is unlikely that any relationship between the two values exists other than a negative correlation, though the correlation again is not strong.

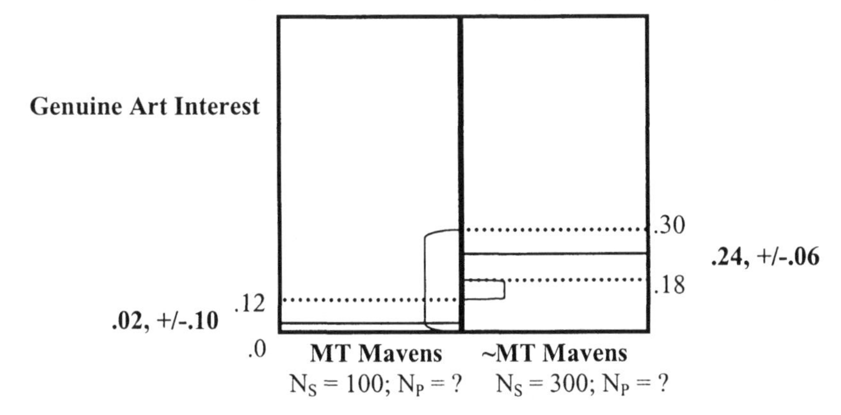

Figure 19.3 Interest in Art: MT Mavens vs. Non-MT Mavens

And so:

A NEGATIVE CORRELATION is a relationship between two variables such that an increase in one variable is accompanied by an decrease in the other and a decrease in one variable is accompanied by an increase in the other.

Difficult Cases and Uncorrelated Variables

What happens, however, when we get overlap? Are we to assume automatically that there is no correlation? No correlation means flatly that two values are not related and this is almost always impossible to state categorically. These questions are not so easily answered and we can perhaps best arrive at answers by offering a couple of illustrations as guides.

Suppose we wish to see whether there are gender-related differences with cigarette smoking. We hypothesize that being female is positively correlated with cigarette smoking—that is, that more women across the globe smoke than do men. We randomly poll 100 men and 100 women across the globe and find that 51 of the men smoke cigarettes, while 45 of the women smoke. Using a ME of +/-.10 for each value, we set up the following diagram.

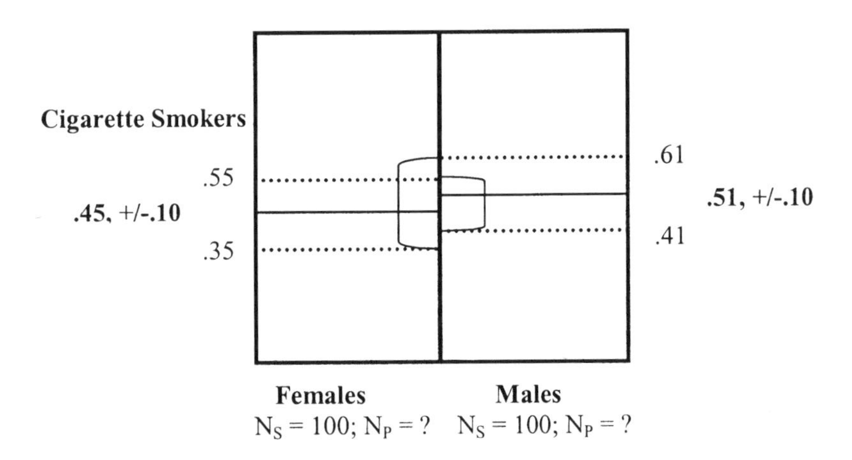

Figure 19.4 Cigarette Smokers$_1$: Females vs. Males

Here we note the range of possible values is from 0.14 to -0.26. This is consistent with a positive correlation between being female and smoking cigarettes (say, if $N_{P(F)}$) = 55% and $N_{P(M)}$) = 42%), a negative correlation (say, if $N_{P(F)}$ = 37% and $N_{P(M)}$ = 58%), or no correlation at all (say, if $N_{P(F)}$ = 44% and $N_{P(M)}$ = 44%). The diagram shows a slight lean toward a negative correlation, but in fact nothing definitive can be said concerning being female and cigarette smoking. Our assessment would thus be that the data are slightly indicative of a negative correlation, but that the data are consistent with a positive correlation and no correlation as well. A larger sample size would help considerably.

Let us finally consider a similar, though slightly different scenario.

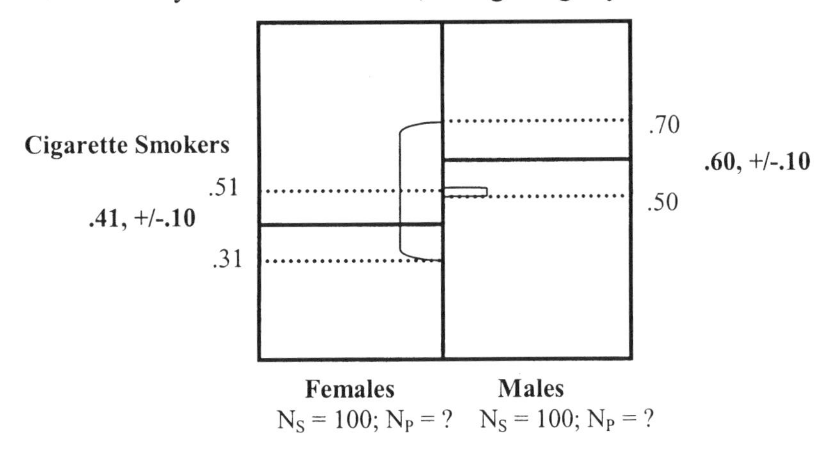

Figure 19.5 Cigarette Smokers$_2$: Females vs. Males

Here, with the range from 0.01 to -0.39, we can pretty much rule out both a positive correlation and no correlation, though both are still remotely possible within the ME of +/-.10. Still, it seems much more likely than not that there is a negative correlation between the two values. Thus, the most cautious approach would be to say that the two are very likely to be negatively correlated, but there is still a very slight possibility of a positive correlation or no correlation at all.

Using the same population as in the example above, let us look at one final example, where the sample size is now very large and the differences between the outcomes is negligible.

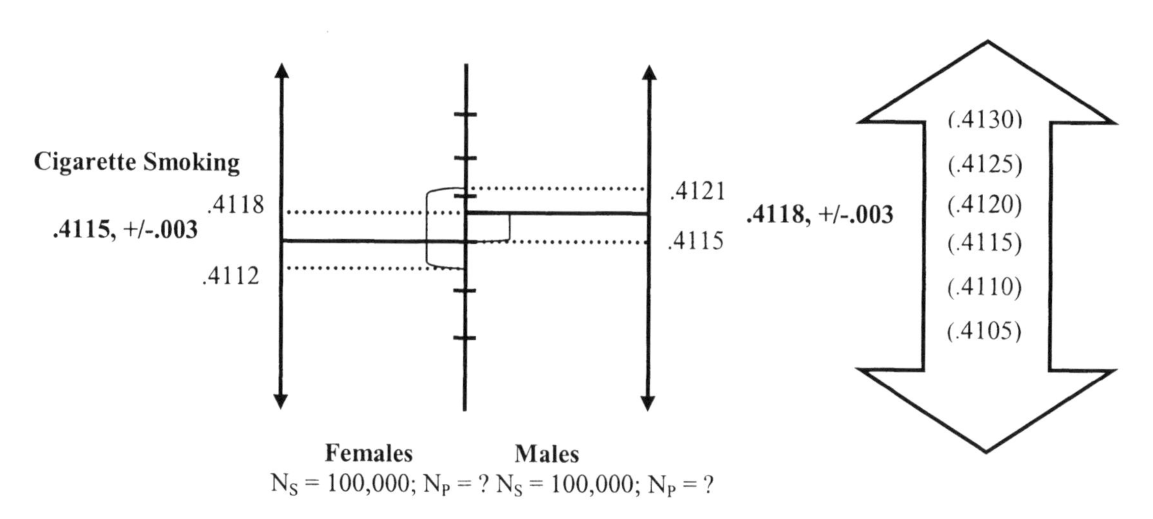

Figure 19.6 Cigarette Smokers$_3$: Females vs. Males

Here, with two such extraordinary large samples and the samples' percentages being so extraordinarily close to each other, 41.15% to 41.18%, no judgment other than "no correlation" can sensibly be given.

And so:

Two variables are UNCORRELATED when a change in one variable is *not* accompanied by a change in the other, and conversely.

A final word. It is important to realize since the relationships we are testing in the examples above are correlational and not causal, we can set up each of the diagrams otherwise. For instance, we can look at samples of smokers and non-smokers (place those variables at the bottom) and see what percentage of those groups is female (or male). The results should be similar—that is, one would find a likelihood of a negative correlation.

KEY TERMS

distribution
negative correlation
positive correlation
no correlation

TEXT QUESTIONS

* What is the difference between a positive and negative correlation?
* What does overlap in the diagrams tell us about the possibility of a correlation or no correlation?

TEXT-BOX QUESTION

Does the 2000 presidential election illustrate, in some sense, that every vote counts?

EXERCISES

Examine the data in the following samples, depicted in the diagrams below. Factoring in how the samples map on to their respective populations, draw the appropriate conclusion for each.

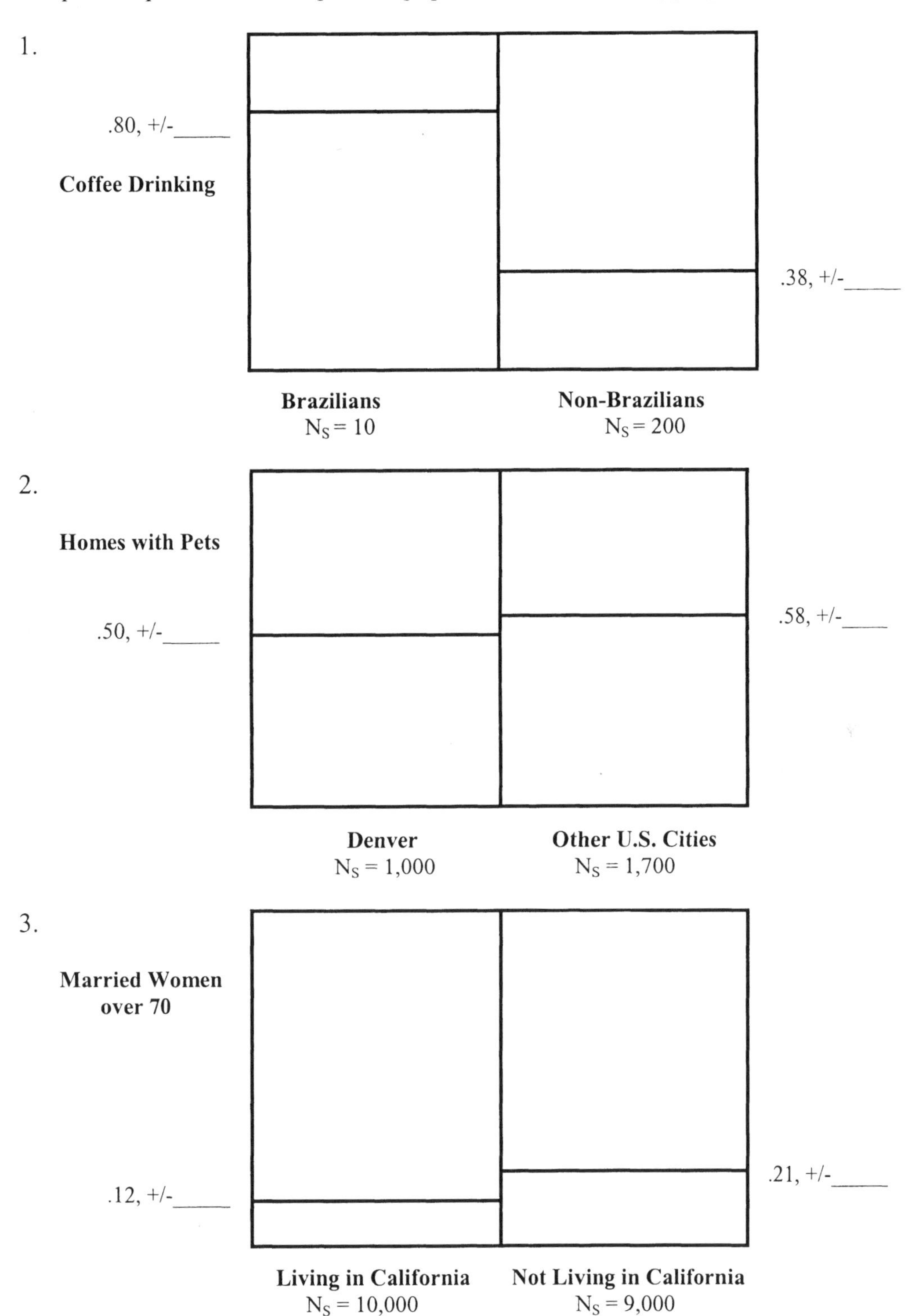

4.

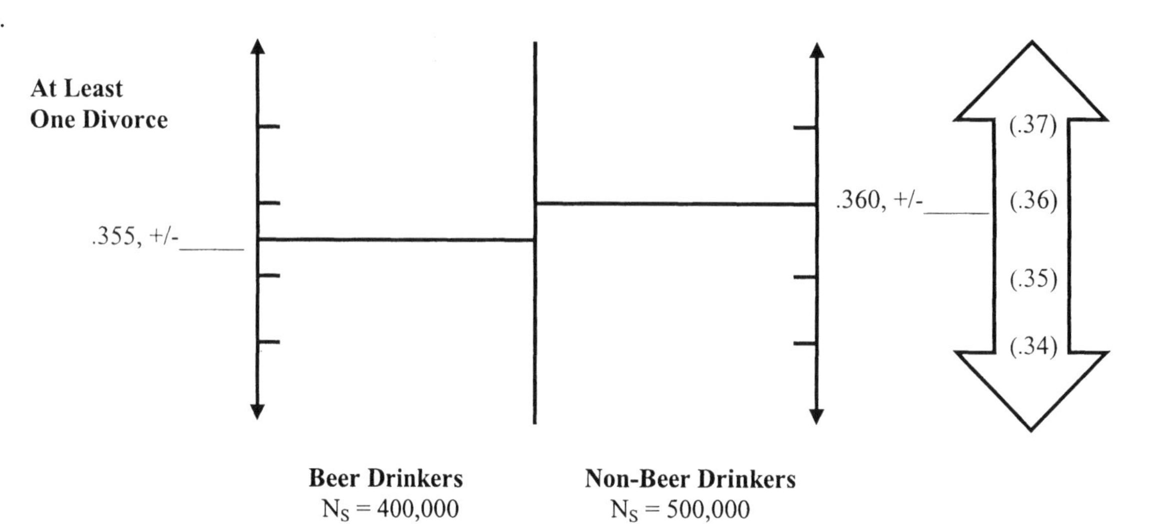
At Least
One Divorce
.355, +/-_____
.360, +/-_____
(.37)
(.36)
(.35)
(.34)
Beer Drinkers
N_S = 400,000
Non-Beer Drinkers
N_S = 500,000

Module 20
Evaluating Statistical Hypotheses

"What I am going to tell you about is what we teach our physics students in the third or fourth year of graduate school... It is my task to convince you *not* to turn away because you don't understand it. You see my physics students don't understand it... That is because *I* don't understand it. Nobody does". Richard P. Feynman, *QED: The Strange Theory of Light and Matter*

THE PREVIOUS TWO MODULES introduced you to three types of statistical models—proportional, distributional, and correlational models. Here, now, are guidelines for evaluating such models.

Guidelines for Evaluation

Evaluating any of the three types of statistical models is merely a matter of tailoring the rules we gave for theoretical models to accommodate the statistical nature of statistical hypotheses.

STATISTICAL HYPOTHESIS: *What is the statistical hypothesis under investigation? Make sure you include where the hypothesis is* proportional, distributional, *or* correlational.

PREDICTION: *What prediction does the statistical hypothesis make? State this in such a way that it straightforwardly suggests the experiment/observations to be done.*

BACKGROUND INFORMATION/RESEARCH METHODS: *Is there background information that is relevant? Is there anything worth mentioning about the research methods? How were the subjects selected?*

RESULTS: *What are the relevant data? Be sure to include the sample size and the margin of error thereby indicated. Draw up any diagrams that may be of help in assessing the data.*

ANALYSIS: *Is it possible to state precisely the intended population?*

If the sample was not random, are there any biases that might make it misrepresentative of the population?

What of the sample's size? Is it large enough to offer a sufficiently good representation of the population? If small, is this a pilot study?

Are the data consistent with the prediction? If so, to what extent do they confirm it? Are there other models or hypotheses consistent with the data that cannot be ruled out? If not, must the hypothesis be rejected categorically or is there some non-ad-hoc way to salvage it?

If the hypothesis is correlational, do the data support a correlation?

CONCLUSION: *Give you verdict here succinctly and precisely.*

Below, let us illustrate by looking at a distributional study and two correlational studies: the first is on the effects of alcohol on next-day memory, the second is from an MSN poll, and the last is on psychological orientation as it relates to performance in sport.

Quantum Strangeness

Statistics are commonly used in science to explain trends in populations, weather patterns, and even birth-rate tendencies. Our use of statistical explanation is such cases, however, does not mean that the phenomena under investigation are not parts of deterministic systems. Statistical explanation, instead, is a reflection of our uncertainty regarding the complex of conditions surrounding such systems, assumed deterministic.

With quantum reality the situation is otherwise. Statistical explanation plays a role in quantum phenomena not because of our uncertainty of lower-level systems, but because such systems seem themselves to be statistical. Again, at the macroscopic level, it is easy to measure the position and velocity of any body at the same time. At the microscopic level, knowledge of the position of any subatomic body comes only through complete loss of knowledge of it velocity; and, conversely, with knowledge of its velocity comes only through complete loss of knowledge of its position. Thus, at the quantum level, explanation is inescapably statistical.

To explain this anomaly, a variety of interpretations have been proposed. I list four.

Neils Bohr's *Copenhagen interpretation* goes as follows. We can never really know what goes on at the quantum level. The very act of investigation determines what we find. An apparatus designed to detect quantum phenomena as waves will find waves. An apparatus designed to detect quantum phenomena as particles will find particles.

Do these seeming acts of "interference" mean that our picture of quantum phenomena is incomplete—that is, that there is more to the picture than we find, when we measure? Bohr did not think so. Rather he thought that quantum particles exist in no definite state *until* they are measured. When measured, they assume the state dictated by the measuring instrument. For Bohr, uncertainty was fundamental to the quantum realm of things; measuring made things certain.

Albert Einstein thought otherwise. Proposing a second, *incompleteness interpretation* of the data, Einstein argued vehemently against the Copenhagen view. He believed that quantum reality was not indeterminate, but completely determinate, just as things at the macro-level. The problem for Einstein was quantum theory itself; the theory was incomplete. Further investigation, he thought, would reveal more about the phenomena under investigation and show that subatomic reality was every bit as deterministic as macro-reality.

Other interpretations have been impacted by more recent investigations about quantum phenomena—specifically *Bell's Interconnectedness Theorem* (1966) and its verification. Specifically, Bell's theorem states that reality is non-local—that is, quantum phenomena can interact with each other over distances at speeds that violate the speed of light as an upper limit of transmission. His theorem has been confirmed through a variety of experiments. In one such experiment, a photon was split into a pair of photons—each travelling opposite the other. Interaction with one of the photons was seen to result in a complementary change in the other in a manner that was simultaneous or, at least, superluminal. Because Bell's theorem, shown correct, gives results that are independent of Quantum Theory, it has imposed further constraints on any interpretation of that theory that add to the strangeness of interpretations of it. I give two further interpretations below that take into consideration Bell's theorem.

A third interpretation of Quantum Theory, an offshoot of the Copenhagen interpretation, was proposed by Eugene Wigner, John von Neumann, and John Wheeler. They argue that the subatomic universe exists, but that subatomic phenomena themselves do not exist until they are observed or measured. It is the act of measurement brings into being quantum phenomena.

A fourth interpretation is the *many-worlds view* of Bryce DeWitt and Hugh Everett. They believe that every attempt at measuring a quantum phenomenon, as when a photon from a measuring instrument interacts with an atom, brings into being a parallel universe with its own history. Particle interactions in this new universe, then, create another universe, and so on.

It seems a good bet that strange interpretations of this strange theory will continue indefinitely.

Next-Day Memory after Alcohol Consumption[1]

Memory Problems after Drink Abuse

Researchers from the University of Utrecht have concluded that the hangover effect of one-night's drinking slows considerably the recovery of memory. This conclusion may not seem too startling, but to date there has been little research in this area.

The researchers studied 48 healthy volunteers—24 men and 24 women—all of whom were considered to be moderate drinkers (i.e., having between 10 and 35 units of alcohol per week), who had experience with a hangover. The subjects were fed a standard meal and then given a memory test. After the test, subjects were randomly split into two groups. The first group was given orange juice, mixed with 1.4 grams of alcohol per kilo of their body weight, which is the equivalent of eight or nine alcoholic beverages. The second group was given orange juice without alcohol. Subjects began drinking their beverages at 11:30 p.m. and were given one-half hour to finish.

The next morning, once their alcohol concentration reached zero (the moment hangovers usually begin), each was given a memory test, where 15 substantive nouns were presented along with other words to distract attention. Subjects were then asked to write down as many of these substantive nouns as they could recall, both immediately afterwards and one hour later. What the researchers observed was that the control group easily outperformed the group given alcohol when asked to recall the nouns one hour later. However, they noticed no significant difference observed for immediate recall.

***Assessment of* Memory Problems after Drink Abuse**

Statistical Hypothesis

Correlational: Consumption of alcohol is *negatively correlated with* recovery of memory recall.

Prediction

If alcohol retards next-day memory, then *subjects who are given tasks involving memory recall will have retarded recall capacities the morning after a day of moderately heavy or heavy drinking.*

Background Information/Research Methods

* The study was done at the University of Utrecht.
* Selection of subjects was not random. All selected were moderate drinkers, who were familiar with a hangover.
* Subjects were 48 healthy, moderate drinkers (10-35 drinks per week), comprising equal numbers of mean and women—24 men and 24 women.
* Each subject was given a standard meal and then a standard memory test. Afterward, they were then split into two random groups.

Results

* Group one: given 1.4 grams of alcohol per kilo of their body weight with orange juice.
* Group two: given pure orange juice. Each group given 30 minutes to drink, beginning at 11:30 p.m.
* The next morning, once their alcohol concentration reached zero (the moment hangovers usually begin), each was given a memory test, where 15 substantive nouns were presented along with other words to distract attention. Subjects were then asked to write down as many of these substantive nouns as they could recall, both immediately afterwards and one hour later.
* Immediate recall: There was no significant different observed between the two groups.
* One hour later: The control group easily outperformed the group given alcohol.
* (Note: I draw up the diagram—Figure 20.1, below—to illustrate the existence of a negative correlation, even though I do not have data.)

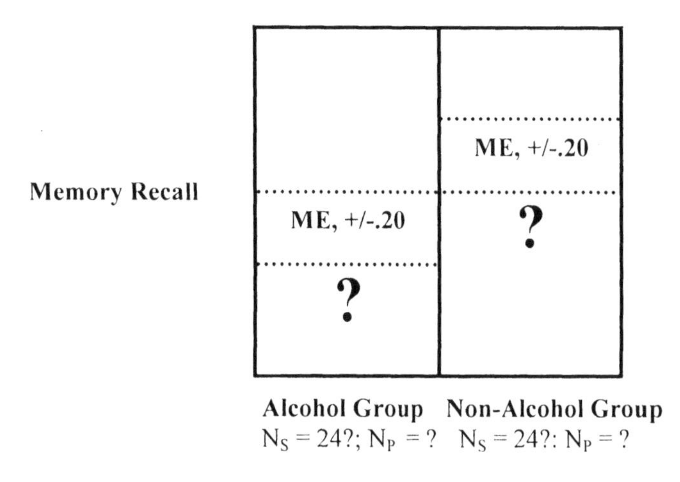

Figure 20.1 Alcohol Consumption and Memory Recall. No data are given here, but the report suggests a negative correlation between alcohol intake and cognitive performance after a one-hour delay. We also assume that that there were 24 subjects (both men and women) in each group.

Analysis

* It is commonly understood that alcohol consumption effects cognitive impairment on the day after consumption. This study is an effort to show just how alcohol impairs cognition. Does it impair immediate memory recall, delayed memory recall or both?
* The data suggest a negative correlation between moderately heavy/heavy consumption of alcohol and retardation on long-term tasks related to memory the next day. The strength of this correlation cannot be determined, as no numbers are included in this report.
* The margin of error is assumed to be +/-.20, on assumption that each of the two groups comprised 24 subjects, with an even mix of males and females.
* The report also suggests that there are additional data from other experiments on the effects of alcohol on cognitive impairment, so this experiment may be merely additional confirmation of a hypothesis that already has support.

Conclusion

The correlation between alcohol consumption and memory retardation for *delayed recall* seems strong. We cannot say more, without data. However, we can refine the hypothesis, adding "delayed recall", given the data of this experiment. We can now say: Consumption of alcohol is *negatively correlated with* recovery of *delayed* memory recall.

Presidential Poll[2]

Should George W. Bush be Impeached?

An MSNBC poll in 2006, asked the following question: "Do you believe President Bush's actions justify impeachment?" Of 310,103 participants, the survey got the following response.

* *86%: Yes, between the secret spying, the deceptions leading to war and more, there is plenty to justify putting him on trial.*
* *4.4%: No, like any president, he has made a few missteps, but nothing approaching "high crimes and misdemeanors".*
* *7.3%: No, the man has done absolutely nothing wrong. Impeachment would just be a political lynching.*
* *1.8%: I don't know.*

***Assessment of* Should George W. Bush be Impeached?**

Statistical Hypothesis

Distributional: Most Americans think that George W. Bush should be impeached.

Prediction

If the hypothesis is true, then *the MSNBC poll will show that the majority of Americans want the president impeached.*

Background Information/Research Methods

* The poll was conducted on MSNBC's home page, under the "Politics" section.
* 310,103 subjects participated.

Results

* 86%: Yes, between the secret spying, the deceptions leading to war and more, there is plenty to justify putting him on trial.

* 4.4%: No, like any president, he has made a few missteps, but nothing approaching "high crimes and misdemeanors".
* 7.3%: No, the man has done absolutely nothing wrong. Impeachment would just be a political lynching.
* 1.8%: I don't know.

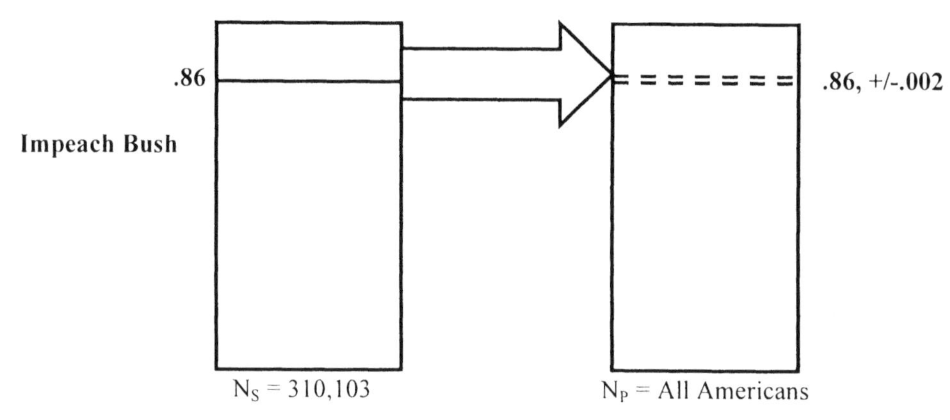

Figure 20.2 U.S. Poll on President Bush

Analysis

The large number of participants make the sample's size sufficiently large to be representative of the intended population, which seems to be all American voters. The ME is +/-.002. However, Bush's overall approval rating at the time of the poll was approximately 39%. In spite of the large sample, the incongruity between Bush's approval rating and the roughly 12% of people who are not in favor of Bush being impeached in the sample is difficult to explain. It is very likely that there is a sampling bias in the poll. We must remember that only those, with access to the internet, who turn to MSNBC as a news source will be replying. It may be that the question was asked in such a manner that mostly those who are very much against Bush would have been much more likely to reply.

Conclusion

It is highly unlikely that the sample is representative of the intended population, in spite of its large size.

Optimizing Athletic Performance through Orientation[3]

Mastery-Orientation and Peak Performance

In study on peak performance of athletes, Susan Jackson and Glyn Roberts hypothesized, among other things, that peak performance would be experienced more by athletes who had a mastery-orientation than those with outcome-orientation. A mastery-orientation is one where one's orientation is perceived to be in the present rather than in the outcome of an athletic event, whereas a high outcome-orientation is the converse.

To test this hypothesis, 200 Division I college athletes from a large mid-western university were studied. There were 110 males and 90 females, who ranged in age from 17 to 25 (mean age being 19.4 years and mean years of involvement in competitive sport being eight years). Freshmen (32%), sophomores (28%), juniors (19%), and seniors (20%) were represented. To avoid potential problems with individual and team goals in one's orientation, only athletes that competed in individual sports were selected—gymnastics (n = 34), swimming (n = 40), golf (n = 21), track athletes (n = 19), cross-country and distance runners (n = 32), field athletes (n = 25), tennis players (n = 20), and divers (n = 9).

The study showed a positive correlation between mastery-orientation and quality of performance. Of 175 athletes that experienced best performance, 116 (66%, +/-.09) described being mastery-oriented. Likewise, only 10 of 144 (7%, +/-.09) reported mastery-orientation during worst performances. Of those reporting best performances, only 24 of 175 (14%, +/-.09) reported an outcome-orientation; of those experiences worst performances, 126 of 144 (88%, +/-.09) reported an outcome-orientation.

Study: Alligators Dangerous No Matter How Drunk You Are (*The Onion*, 42.19, May 10, 2006)

Our data strongly indicates that human intoxication does not transform an alligator into a docile creature that enjoys wrestling", said professor Ryder McCrory, chair of the Wildlife Taunting Department of LSU's prestigious Center For Bullying And Hazing Studies. "Despite its slow-witted demeanor and tendency to bask motionlessly in the hot sun, it's a mistake to believe that an alligator will passively tolerate a half nelson, no matter how much Southern Comfort is fueling it". McCrory said the study yielded statistics that speak for themselves. "In 10 out of 10 documented cases of violent alligator–drunkard encounters, the reptile was not influenced by the fact that the victim was 'just kidding' or 'just having some fun'", McCrory said. To an alligator, McCrory explained, a human forearm, even drunkenly dangled between the creature's casually opened jaws, still appears to be prey.

In a breakthrough study that contradicts decades of understanding about the nature of alligator–drunkard relations, Louisiana State University researchers have concluded that people's drunkenness does not impair the ancient reptiles' ability to inflict enormous physical harm. "Alligators exhibit the potential to inflict serious harm, regardless of the blood-alcohol levels of their victims".

In field experiments, members of the control group performed no better—and often far worse—than their sober counterparts in defending themselves against a 300-pound, seven-foot bull alligator. Even when armed with an empty tequila bottle." At best, the bottles bounced harmlessly off the alligator's snout", said LSU research assistant Tracy Sawyer. When placed in water, the drunken volunteers fared even worse, and the alligator markedly better, Sawyer said. In addition, the alligators far outperformed their inebriated human counterparts in the following areas: lunging, biting, crushing, dismembering, and swallowing.

According to the study, an alligator's characteristic grin should not be interpreted as a lighthearted reaction to the outrageous nerve of an alcohol-addled human. "Don't let an alligator's easygoing appearance fool you", Sawyer said. "These creatures have no empathy for drunken pranksters looking for fun. They are not black bears".

McCrory recommended that alligator wrestling be undertaken solely by professionals, specifically roadside-attraction proprietors. For drunkards interested in proving their mettle with alligators, the researchers proposed these guidelinesInstead of baiting an alligator, seek another form of drunken recreation, such as attending a strip club, burning a pile of tires, or painting one's buttocks with a funny face and videotaping it. Sick or infant deer are considered a far safer match for most inebriated humans; kicking a raccoon or squirrel already dying by the side of the road is also recommended.

Experts suggest that those who become aggressive after consuming alcohol would be safer channeling that energy into more constructive behavior, such as calling an ex-lover. And McCrory warned drunkards who "absolutely must assault an alligator while inebriated" to first make sure it is not a John Deere Gator cargo utility vehicle. This oversight "is a common occurrence", he said.

The researchers, however, advise caution before a hasty interpretation of their results: "Due to the associational nature of this study, directionality cannot be assumed; it is quite possible that performing successfully in a competitive situation induces a positive psychological state. Further research is needed to clarify the nature of the relationship between the perspective held in competition and one's quality of performance".

Assessment of Mastery-Orientation and Peak Performance

Statistical Hypothesis

Correlational: Peak athletic performance *is positively correlated with* mastery-orientation.

Prediction

If the hypothesis is true, then *athletes with a mastery-orientation will experience peak performance more than athletes with an outcome-orientation.*

Background Information/Research Methods

* To avoid problems with individual and team goals in an athlete's orientation, only athletes from individual sports were selected.
* 200 Division I college athletes of a large mid-western university. There were 110 males and 90 females, ranging in age from 17 to 25 (mean age, 19.4 years, and mean years of involvement in competitive sport, eight years). Freshmen (32%), sophomores (28%), juniors (19%), and seniors (20%) were each well represented.
* Sports represented: gymnastics (n = 34), swimming (n = 40), golf (n = 21), track athletes (n = 19), cross-country and distance runners (n = 32), field athletes (n = 25), tennis players (n = 20), and divers (n = 9).

Results

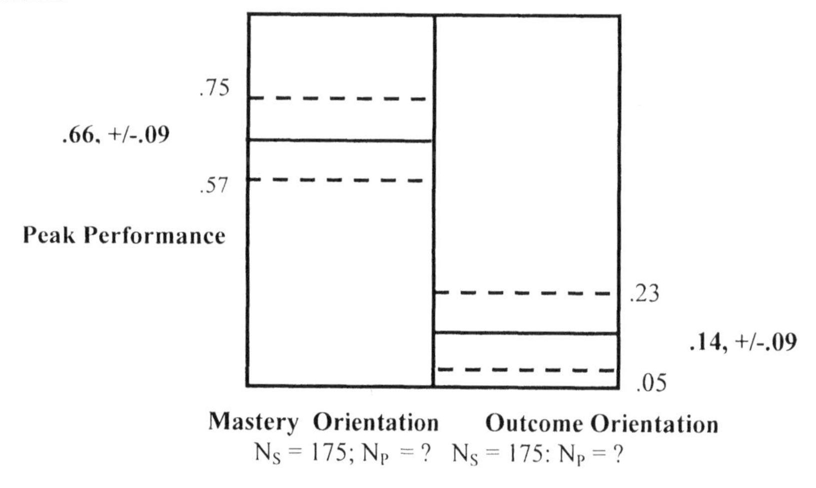

Figure 20.3 Peak Performance: Mastery-Orientation vs. Outcome-Orientation.

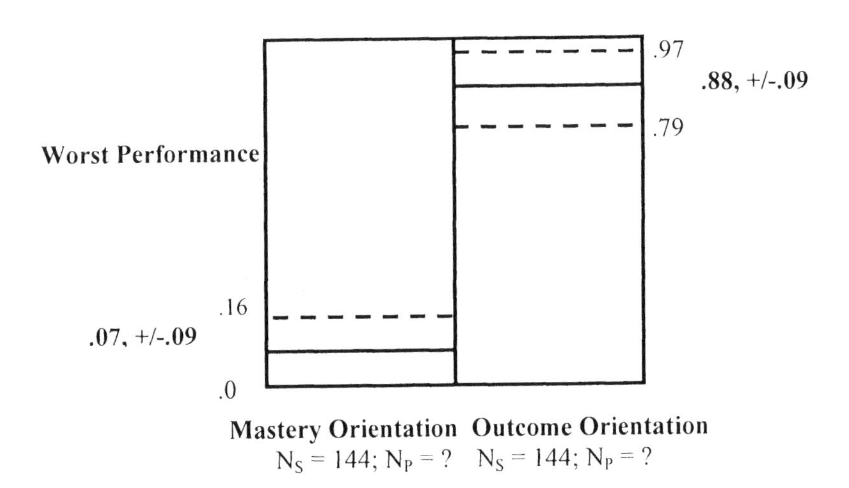

Figure 20.4 Worst Performance: Mastery-Orientation vs. Outcome-Orientation

* 116 of 175 (.66, +/-.09) athletes described best performance as being mastery-oriented. 24 of 175 (14%, +/-.09) of those with best performance reported outcome-orientation (Figure 20.2, above).
* 10 of 144 (.07, +/-.09) reported mastery-orientation during worst performances; 126 of 144 (88%, +/-.09) with worst performance reported outcome-orientation (Figure 20.3, above).

Analysis

* There is a positive correlation between mastery-orientation and peak performance, the strength of which ranges between .34 and .70 (with the .95 confidence measure).
* There is a negative correlation between competitive orientation and peak performance, the strength of which ranges between -.63 and -.97 (with the .95 confidence measure).
* The correlations seem undeniable. However, nothing causal can be said at this point, as it is possible that mastery-orientation may be a cause of peak performance. Likewise, it is also possible that peak performance may be a cause of mastery-orientation.
* (Note the variables are reversible, as the hypothesis is correlational, not causal. One could treat performance as the independent variable [labeled at the bottom] and outcome as the dependent variable [labeled at the side, left]. If mastery-orientation is *causally* related to peak performance, it would be necessary to run additional experiments where the orientations are the independent variables, to be placed at the bottom of the diagrams and the extremes of performance are the dependent variables, as we have done. For correlational models, where variables are placed is irrelevant, as we are only testing to see whether they are related, not whether they are *causally* related.)

Conclusion

There is a positive correlation between peak performance and mastery-orientation and a negative correlation between competitive orientation and peak performance. Further research should to be done to show directionality (i.e., which variable is the cause and which variable is the effect).

TEXT QUESTIONS

* In what ways do the guidelines for evaluating statistical hypotheses differ from those of theoretical hypotheses?
* How does the study on alcohol consumption and memory loss show that researchers are often forced by data to refine their hypotheses?
* Why must one always be cautious in evaluating statistical claims? Use the study on peak athletic performance and psychological orientation as a guide here.

TEXT-BOX QUESTION

What are the four different interpretations of Quantum Theory listed in the text box? Which interpretation makes the most sense to you, or do you, like Richard P. Feynman (see introductory quote), think that the theory is incomprehensible?

---

1 Joris C. Verster, Danielle van Duin, Edmund R. Volkerts, Antonia H.C.M.L. Schreuder, and Marinus N Verbaten, "Alcohol Hangover Effects on Memory Functioning and Vigilance Performance after an Evening of Binge Drinking", *Neuropsychopharmacology* (2003) 28, 740-746.
2 http://www.msnbc.msn.com/id/10562904/.
3 Susan Jackson and Glyn Roberts, "Positive Performance States of Athletes: Toward a Conceptual Understanding of Peak Performance", *The Sport Psychologist,* 1992, 6.

Module 21
Exercises for Statistical Hypotheses

"A low voter turnout is an indication of fewer people going to the polls". George W. Bush

EVALUATE EACH OF THE FOLLOWING EPISODES according to the guidelines summed below.

STATISTICAL HYPOTHESIS: *What is the statistical hypothesis under investigation? Make sure you include where the hypothesis is* proportional, distributional, *or* correlational.
PREDICTION: *What prediction does the statistical hypothesis make? State this in such a way that it straightforwardly suggests the experiment/observations to be done.*
BACKGROUND INFORMATION/RESEARCH METHODS: *Is there background information that is relevant? Is there anything worth mentioning about the research methods? How were the subjects selected?*
RESULTS: *What are the relevant data? Be sure to include the sample size and the margin of error thereby indicated. Draw up any diagrams that may be of help in assessing the data.*
ANALYSIS: *Is it possible to state precisely the intended population?*
If the sample was not random, are there any biases that might make it misrepresentative of the population?
What of the sample's size? Is it large enough to offer a sufficiently good representation of the population? If small, is this a pilot study?
Are the data consistent with the prediction? If so, to what extent do they confirm it? Are there other models or hypotheses consistent with the data that cannot be ruled out? If not, must the hypothesis be rejected categorically or is there some non-ad-hoc way to salvage it?
If the hypothesis is correlational, do the data support a correlation?
CONCLUSION: *Give you verdict here succinctly and precisely.*

Like other examples of scientific reports you will come across, some of these may not allow for a complete evaluation at each of the six steps listed above. Do the best you can to fill in the steps.

The Elderly and Sleep[1]

Sleep: A Question of Quantity or Quality?

Research by psychologist Sonia Ancoli-Israel and colleagues suggests that the elderly's drowsiness, inattentiveness, daytime fatigue, and daytime napping may not be due to the perception that people need more sleep as they age. They studied sleep patterns of 1000 people, randomly picked from a San Diego telephone directory. Of these, 427 were 65 or older, while 544 who were under 65. Information about each person's diet, exercise habits, and daytime routine was gathered. Then, all subjects had their breathing, muscle tension, and time asleep monitored during one-night's sleep at home. The morning after, each was asked to fill out a questionnaire de-

scribing how they slept.

Differences between the two groups were noticeable. Most of those in the 65-or-older group experienced nighttime disturbances. Forty-eight percent suffered from either sleep apnea (where breathing periodically ceases during sleep) or nocturnal myoclonus (where sleepers uncontrollably kick and jerk legs in sleep). Ten percent had both. Of the under-65 group, only 15 percent reported either disturbance, while three percent had both.

These types of disturbances may be the real causes of lethargy among the elderly during waking hours. William Dement, director of the Sleep Disorders Center at the Stanford University School of Medicine, states that the sleep of a healthy 25-year-old person to that of an elderly person is radically different. Over a seven-hour period, an elderly person will wake some 153 times, while a 25 year old will wake some 10 times.

Albert Einstein (1879-1955)

Widely regarded as the most significant scientist of the 20th century, Albert Einstein was born in Württemberg, Germany on March 14 of 1879. When his family moved to Italy, Einstein entered the Swiss Federal Polytechnic School in Zurich and studied physics and mathematics. After earning his diploma in 1901, he could not find a teaching position, so he took a job as technical assistant in the Swiss Patent Office, till he earned his Ph.D. in 1905.

It was at the patent office perhaps that he did his most remarkable work. In 1905, he wrote four papers—three of which were groundbreaking (Brownian Motion, the Photoelectric Effect, and Special Theory of Relativity). In 1909 he became Professor Extraordinary at Zurich. Two years later, he became Professor of Theoretical Physics at Prague in 1911.

In 1914, Einstein moved to Berlin, where he was appointed Director of the Kaiser Wilhelm Physical Institute and Professor at the University of Berlin. In 1915, he gave a series of lectures at the Prussian Academy of Science, where he put forth a new theory of gravity. The General Theory of Relativity was born.

Einstein was to remain in Berlin until 1933, when he renounced his citizenship because of Hitler's rise to power in Germany. In 1933, Hitler passed "The Law of the Restoration of the Civil Service", which forced all Jewish university professors out of their jobs. Einstein was smeared by the Nazis and his work was called "Jewish physics". German professors who taught Einstein's theories were blacklisted. Noteable among these was Werner Heisenberg—discoverer of the "Uncertainty Principle" of Quantum Mechanics (see tetboxes, Modules 18 and 20). Einstein renounced his Prussian citizenship and accepted a position as Professor of Theoretical Physics at Princeton University in their newly founded Institute for Advanced Study. Still retaining his Swiss citizenship, Einstein became an American citizen in 1940. He remained at Princeton till his retirement.

Of his scientific views he said:

> I believe that every true theorist is a kind of tamed metaphysicist, no matter how pure a "positivist" he may fancy himself. The metaphysicist believes that the logically simple is also the real. The tamed metaphysicist believes that not all that is logically simple is embodied in experienced reality, but that the totality of all sensory experience can be "comprehended" on the basis of a conceptual system built on premises of great simplicity.

He died of heart failure on April 18 of 1955. At the time of his death, he was one of the most recognizable persons in the world.

Just One More Incident of Brutality in Hockey

MSNBC Poll on Fighting in Hockey

On March 8, 2004, Vancouver Canucks forward Todd Bertuzzi punched Colorado Avalanche forward Steve Moore, himself an "enforcer", and sent him face-first to the ice. Moore suffered a broken neck, deep facial cuts, and a concussion. The attack came in retaliation for a hit Moore put on Markus Naslund, the Vancouver captain and NHL scoring leader at the time, that was not penalized and put Naslund out of action for three games.

Two days later, MSNBC Sports polled its readers on fighting in hockey. They asked, "What should the NHL do about fighting?" Of 2515 responses, the responses were as follows: 52% said "ban it", 28% said "keep rules the same", and 20% said "make it easier to fight back".

Do Beer Drinkers Really Have Bigger Bellies? [2]

Beer and Beer Bellies

Research conducted by scientists in Britain and the Czech Republic suggests that there is no such thing as a beer belly. These scientists surveyed nearly 2,000 Czechs, generally considered to be the world's biggest beer drinkers, as they drink more beer per person than people in any other country. Writing in the European Journal of Clinical Nutrition, *they claimed that there is no link between beer consumption and the size of one's belly.*

Martin Bobak of the University College London and colleagues at the Institute of Clinical and Experimental Medicine in Prague questioned 891 Czech men and 1,098 women between the ages of 25 and 64. Subjects were selected randomly among those who drank either no alcohol or only beer. Men in the study drank on average 3.1 liters of beer per week while women drank on average 0.3 liters per week. Only 3% of the men drank more than 14 liters of beer in a week and just five women drank more than 7 liters in a week, so few of the subjects were considered heavy drinkers. Subjects also underwent a short medical examination. Doctors measured their weight and their waist-to-hip ratio and body-mass index as measures of their obesity.

The scientists found no link between beer drinking and obesity. They said there is a common misperception that drinking beer gives one a beer belly., "There is a common notion that beer drinkers are, on average, more 'obese' than either non-drinkers or drinkers of wine or spirits. This is reflected, for example, by the expression 'beer belly'. If this is so, then beer intake should be associated with some general measure of obesity, such as body mass index or with indices of fat distribution such as waist to hip ratio or with both". Yet they added, "The association between beer and obesity, if it exists, is probably weak".

British Dietetic Association's Nigel Denby warned that beer lovers should not rush to the local pub. "People shouldn't assume that they can now drink freely", he told BBC News Online. "Any food taken in excess can lead to obesity. Drinking any type of alcohol can also lead to obesity. People who want to drink should enjoy alcohol but ... in moderation".

Do Advice Clinics Contribute to Sexually Transmitted Infections?[3]

Advice Clinics and STIs

New research suggests that family-planning clinics encourage the spread of sexually transmitted diseases. In areas where contraception and advice on sex are made more widely available, rates of sexually transmitted infections actually rise.

Between 1999 and 2001, sexually transmitted infections among teenagers increased by around 15%, while the number of adolescent family-planning clinic sessions rose from 27,075 to 33,369—an increase of 23.2%. David Paton, head of industrial economics at Nottingham University, has this to say:

> *Across the country, the estimates imply that the increase in clinic sessions between 1999 and 2001 caused sexually transmitted infection rates to increase by 1.42%.*
>
> *We are pretty sure, statistically, that the increase in clinic sessions was associated with an increase in diagnosis rates. Sexually transmitted infection (STI) diagnoses increased by about 15% between 1999 and 2000—one tenth of this was due to clinics.*
>
> *If clinic sessions had the impact of increasing STIs, even by a small amount, and did not decrease teenage pregnancy rates, then the money and resources would have been better spent elsewhere. Indeed, even doing nothing would have been a better tactic.*

Paton's research concerned 99 health authorities in England. He compared the number of teenage family-planning clinic sessions with levels of sexually transmitted disease and found a correlation between the two. In one inner-city area, the number of clinic sessions increased by 81% and the diagnosis rate of sexually transmitted diseases for those under 20 rose by 70%. In a commuter-belt area, clinics went up by 26% while sexually transmitted diseases also went up by 26%. Paton said:

> *When we promote condoms, we think that might have the effect of preventing passing on disease but it may encourage people to have more partners. Distributing condoms offers protection in preventing disease but people change their behavior and tend to take more risks. It seems to me that setting up sexual health clinics is easy. Social exclusion and deprivation are more difficult to deal with.*

In 1999, the year in which the teenage pregnancy strategy was introduced in England, 61% of 16 to 19 year-olds were sexually active. This increased to 67% in 2000 and 73% in 2001. Over the same period, conception rates among all teenagers fell by 3.5% while rates of sexually transmitted infections rose by 15.8%. The report, which will be published in an economics journal, concludes: "The provision of family planning services for young people appears to have little overall impact on teenage pregnancy rates, but leads to significantly higher rates of diagnosis of sexually transmitted diseases among teenagers".

Paton maintains that teenagers face pressures from peers and magazines to have sex. In addition, the erosion of religious and social taboos about underage sex has contributed to greater sexual activity and higher levels of disease.

Uganda has had considerable success in fighting HIV. It has adopted what it calls the "ABC Approach" to reducing HIV: "Abstain (from sex), Be Faithful (together), Condom Use (every time)". HIV rates in the country have reduced from 21% to 9.8% from 1991 to 1998.

Daniel Low-Beer of the Health and Population Evaluation Unit at Cambridge University believes that Uganda has successfully reduced sexually transmitted diseases, as it has reduced casual sex, instead of focusing on the Western safe-sex approach based on wider access to contraception. Low-Beer says: "We don't do what we say in the UK. The safe-sex policy has not been safe in the UK and, when exported to Africa, it has become quite dangerous. In the last 10 years most countries that have followed the safe-sex approach have not seen a decline in HIV".

A study by Low-Beer and Rand Stoneburner, in the African Journal of Aids Research, *found that developing countries succeed in reducing HIV, when they focus on reducing casual sex. The study found that, between 1989 and 1995, casual sex in Uganda declined by 65% and Low-Beer believes the UK could learn from Uganda's success. States Stoneburner: "We have got to look at the reality of safe sex in the UK. There is inconsistent condom use and an increase in casual sex. The context is very different in the UK but we need to listen to what has happened in Uganda. We need to put risk avoidance into the sexual health strategy. Avoidance is the reduction in casual sex and, in some cases, abstinence and considered condom use".*

The Family Planning Association (FPA) disputed the findings of Paton and Low-Beer. "As far as the FPA is concerned, the message is that one way to reduce STIs is to have fewer partners but the important thing is not what you do but the way that you do it. There are many ways to have sex safely. The most important element is careful and consistent condom use".

STIs have been on the rise since the early 1990s. This is part of an ongoing trend, it did not start in 1998 as Dr Paton is suggesting. There are not enough clinics for young people who want access to them. The rise in STIs is due to a multiplicity of factors including poor sex and relationship education, a lack of health education campaigns and greater diagnosis. Just to link the rise in STIs to an increase in clinics is a weak argument.

Ineffective Antidepressant?[4]

Paxil and Suicide?

GlaxoSmithKline and the Food and Drug Administration warned in a letter to doctors that the antidepressant Paxil may raise the risk of suicidal behavior in young adults. The letter was accompanied by changes to the labeling of both Paxil and Paxil CR, a controlled-release version of the drug, also called paroxetine.

The letter stated that recent clinical-trial data on nearly 15,000 patients treated with Paxil and dummy pills showed a higher frequency of suicidal behavior in young adults treated with Paxil. There were 11 suicide attempts—none resulting in death—among the patients given Paxil in the trials, while only one of the dummy-pill patients attempted suicide. Given that small number of suicide attempts, the results "should be interpreted with caution", the FDA said. Eight of the 11 attempts were patients between the ages of 18 and 30. All patients suffered from psychiatric disorders, including major depression.

GlaxoSmithKline released its findings following an FDA request that antidepressant manufacturers examine their clinical trial data for any links between the drugs and suicide in adults. Company spokeswoman Mary Anne Rhyne said, "At some point, the FDA is going to say what their analysis shows across the category". She added, "We felt like this was information we wanted to share with physicians".

John E. Kraus, the company's director of clinical development for clinical psychiatry in North America, said in the letter that GlaxoSmithKline continues to believe that Paxil's benefits outweigh its risks. In the same letter, the FDA stressed that there needs however to be careful monitoring of all patients taking Paxil—especially young adults and those who are improving.

Assessment of Sleep: A Question of Quantity or Quality

Statistical Hypothesis

Prediction

Background Information/Research Methods

Results

Analysis

Conclusion

***Assessment of* MSNBC Poll on Fighting in Hockey**

Statistical Hypothesis

Prediction

Background Information/Research Methods

Results

Analysis

Conclusion

Assessment of Beer and Beer Bellies

Statistical Hypothesis

Prediction

Background Information/Research Methods

Results

Analysis

Conclusion

Assessment of Advice Clinics and STIs

Statistical Hypothesis

Prediction

Background Information/Research Methods

Results

Analysis

Conclusion

Assessment of Is Paxil Linked with Suicide?

Statistical Hypothesis

Prediction

Background Information/Research Methods

Results

Analysis

Conclusion

Text Questions

* In the poll on fighting in hockey, what sorts of sampling biases are present?
* What sort of causal fallacy (see Module 9) is assumed in the report on advice clinics and STIs?

Text-Box Questions

* Why are possibility claims often times very misleading as conclusions of studies? What are two proper uses of such claims in studies?
* Why are necessity claims always illegitimate in science?
* What is the rule of thumb for evaluating all claims involving neither necessity nor possibility?
* Why are polls generally more reliable than online surveys?

---

1 Joshua Foschman, "Golden Years and Restless Nights", *Psychology Today*, February 01, 1986.
2 "Why the Beer Belly May Be a Myth", *BBC News, UK Edition,* Oct. 12, 2003, http://news.bbc.co.uk/1/hi/health/3175488.stm.
3 Sarah-Kate Templeton, "Advice Clinics 'Increase Sexual Diseases'", *Sunday Herald Online*, Nov. 2, 2003, http://www.sundayherald.com/37766.
4 "FDA Warns of Suicide Risk for Paxil", *Associated Press*, May 12, 2006, http://news.yahoo.com/s/ap/20060512/ap_on_he_me/paxil_suicide_risk.

SECTION SEVEN

Causal Hypotheses

Module 22
Controlled Experiments & Data-Based Research

"Self-determination does not follow from complexity". B.F. Skinner

THE PREVIOUS SECTION ANALYZED statistical hypotheses. We saw that such hypotheses are tremendously important in scientific experimental research—especially in its earliest stages. Not only do they drive preliminary research, they also suggest new directions for future research, through a search for causal relationships between variables.

The aim of this section is to offer a brief introduction to controlled scientific experiments as an introduction to causal hypotheses. The first part of this module gives a brief explanation of controlled experiments. Next, I present the three different kinds of controlled experiment—randomized, prospective, and retrospective—and evaluate the relative strengths and weaknesses of each approach. Finally, I introduce a relatively recent research technique—data-based research.

What Is a Controlled Experiment?

The *controlled experiment*, unlike haphazard observation, is a systematic attempt to confirm a hypothesis by gathering information in support of it in a controlled manner. Such an experiment, when it is completed, may be viewed as an inductive generalization that attempts to establish a causal link between phenomena. Let us consider the following hypothesis.

H: Pre-surgical stress causes post-surgical problems.

This hypothesis seems quite sensible. Stress, in general, is known to be bad for bodily health, so it seems reasonable that stress prior to surgery could be a cause of post-surgical complications.

How can we test the causal hypothesis? First, it is important to note that the hypothesis is a statement that has scope for all surgical situations—past, present, and future. When we *test* for it, however, we must limit ourselves to a certain finite number of observations. How many? There is, as you may have guessed, no set number. Of course, more observations make it more likely that the hypothesis will be true of the intended population—here, all surgical situations. Time and money are always issues. In this scenario, if there is no prior research, then a relatively small study is reasonable to begin. If the data prove not to be significant, few resources will have been lost.

To set up a test for the causal hypothesis, a researcher could begin, say, by interviewing patients prior to a particular type of stressful surgery and breaking up all those who are interviewed, given the data of the interview, into two groups: high pre-surgical stress (experimental group) and low pre-surgical stress (control group). The researcher could then monitor the two groups for the number and kinds of post-surgical problems that arise.

Not surprisingly, a review of the medical literature offers confirmation of the hypothesis. However, there is a catch. These studies have tended to focus on long-term post-surgical problems—that is, how stress impacts the immune system weeks and even months after surgery. Yet some studies on the short-term post-surgical impact of stress suggest that those who go into and through surgery in a relatively stress-free manner suffer weight loss, fatigue, and impaired immune functioning much more so that those who are highly stressed. Those complications may arise, because relaxation seems to spark two key potentially dangerous stress hormones. Thus, anxiety before surgery may actually be beneficial, not harmful, in the short term.[1] Overall, then, the hypothesis, as it stands, is ambiguous because "post-surgical stress" does not distinguish between short-term and long-term stress. More research, it is likely, needs to be done to clear up the confusion. Such research could be fueled by two hypotheses:

H_{LT}: Pre-surgical stress causes long-term post-surgical problems.
H_{ST}: Pre-surgical stress prevents short-term post-surgical problems.

Types of Controlled Experiments

There are three types of controlled experiments: randomized experiments, prospective experiments and retrospective experiments. I examine each, in turn, below.

Randomized Controlled Experiment

A *randomized controlled experiment* is so-called because subjects are selected randomly from a population to represent that population. Let us begin with a definition of *random sample*.

A sample is RANDOM if and only if each member of a targeted population has exactly the same chance of being selected.

In practice, perfectly random samples are almost always impossible, so researchers look at randomization as an ideal to be approximated as nearly as possible.

Once selected for a randomized experiment, subjects are then broken into two groups—the control group (G_C) and the experimental group (G_E). The control group provides a basis for comparison, as only the experimental group is really being tested.

In a RANDOMIZED CONTROLLED EXPERIMENT of a suspected causal relationship between two variables α and ω (where α is the suspected cause and ω is the suspected effect), subjects in G_E are exposed to α, while subjects in G_C are not (or given a placebo). If a statistically significant difference is observed between those in G_E that come to have ω and those in G_C that come to have ω, then there is reason to believe α is a cause of ω.

In the best experiments, neither subjects nor researchers know which subjects are in which group. Otherwise, there is the possibility of experimental bias, known as the *placebo effect*. Consider the hypothesis, "Small amounts of caffeine prior to exercise delays muscular fatigue". Now suppose, in testing for that, we should tell subjects in the control group that their beverage has no caffeine in it and subjects in the experimental group that their beverage has caffeine in it. Subjects who are told they are given caffeine prior to exercise may expect improved performance, as this is the hypothesis under investigation, so they will push themselves harder than others told beforehand that they will be given a placebo. Yet even if subjects are blinded, researchers might be inclined to interact with subjects in such a way to bias results or to interpret data favorably, if they are not blinded also. After all, successful research often means prestige for scientists and research institutions. In consequence, in the best studies, both subjects and researchers are blinded. In short:

A DOUBLE-BLIND STUDY is one in which neither subjects nor researchers can tell G_E from G_C.

A BLIND STUDY is one in which only subjects cannot tell G_E from G_C.

As randomized controlled experiments are generally done in laboratories, the number of subjects that can be studied is usually small, but the controls that can be imposed under laboratory conditions are many. For instance, suppose we wished to study the impact of red wine on heart disease. We would select two groups of subjects with arterial constriction. We might give each group the same diet and the same amount of daily exercise, and so forth—the idea being to keep the conditions as similar as possible for each subject in both groups. Next, we might give the experimental group two glasses of red wine each day for a certain length of time and fail to do so for the control group. Overall, these controls would put researchers in a good position to claim that any observed difference in the two groups (e.g., freer blood flow in the experimental group) is due to the red wine.

Imagine performing a similar experiment outside of a laboratory. Researchers would have little control over what their subjects ate, their exercise patterns, and the use of other nutrients or drugs that might impact heart disease in one way or another. Moreover, imagine also that these stay-at-home subjects were volunteers. It seems plausible that those volunteering for such a study would be people who like red wine and who probably drink two or more glasses each day anyway. What also will prevent subjects from restricting wine intake to two glasses? What assurance could experimenters have that those in the control group are *not* drinking red wine? Everything depends on the sincerity of the subjects in their verbal reports.

What are the drawbacks of randomized controlled experiments? Although laboratories allow for numerous controls, they also introduce certain difficulties. I list three below.

First, laboratories are artificial environments and researchers can never be sure that results observed in a laboratory will obtain outside of the laboratory. For instance, studies of the content of dreams that are done in sleep labs suffer from one serious flaw: Sleep environments impact the content of dreams and most peoples' sleep environment is relatively stable. Why should one have the same dreams in a laboratory that one has in one's own bed?

Second, laboratory studies, especially those that are lengthy, are generally expensive. Thus, to see whether there is a causal relationship between two variables, researchers tend to speed up the results in ways that call into question the very claims made. For example, in studies on suspected carcinogens, say a particular spray used only on Michigan apples, laboratory rats may be given a ridiculously large amount of the suspected carcinogen to speed up the results. If a noticeable percentage of the rats in the experimental group develop cancer, the most appropriate

Explaining the Placebo Effect

The placebo effect is the measurable or perceived improvement in health that is not attributable to treatment. A placebo (Latin, "I shall please") may take the form of a medicine administered to a patient or any form of treatment given to a patient that is thought to be causally ineffective in bringing about a desired effect.

There are several different theories that have been put forth to explain the placebo effect. Perhaps the most widely held view concerns the role of expectation. Subjects who are given placebos fully expect the medicine or treatment to have some effect and so they perceive some effect. Psychologists Irving Kirsch and Guy Sapirstein have done considerable analysis of data relating to the effects of antidepressants and the placebo effect. They concluded 75 percent of the effectiveness of antidepressants is due to patients' expectation of improvement, not the power of the drugs administered. Kirsch says, "The critical factor is our beliefs about what's going to happen to us. You don't have to rely on drugs to see profound transformation". What Sapirstein and Kirsch have found is that placebos can actually have significant biochemical effects on subjects. They conclude that thinking and sensing have an affect on biochemistry.

A second explanation for the placebo effect is simply time. Many illnesses and injuries can be sufficiently healed spontaneously over time, without the intervention of any medicine or treatment. The problem with this approach is simply that people given placebos seem to do better, both in the short and long term, than those given no treatment at all. Writes psychiatrist Walter A. Brown of Brown University:

> There is certainly data that suggest that just being in the healing situation accomplishes something. Depressed patients who are merely put on a waiting list for treatment do not do as well as those given placebos. And—this is very telling, I think—when placebos are given for pain management, the course of pain relief follows what you would get with an active drug. The peak relief comes about an hour after it's administered, as it does with the real drug, and so on. If placebo analgesia was the equivalent of giving nothing, you'd expect a more random pattern.

A third account of the placebo effect is that it not the placebos given to patients that trigger improvement, but instead the care patients are shown in treatment. Many now believe that caring physicians and attendants—through touching, encouragement, and personalized attention—matter as much as the medicines and treatment given. Such care, in turn, may trigger physical changes, such as the release of endorphins, which reduce overall stress and promote healing.

Then again, the answer might be found in stress, as we indicate earlier in this module. Patients who are given a placebo are likely to feel less stressed in the anticipation that help is on the way, so to speak. Diminished stress may then be causally responsible for improved immunological functioning over time.

Whatever the correct explanation—and the four accounts sketched above may all be part of the explanation—the placebo effect is the key to explaining the reported effects of certain questionable alternative practices like homeopathy and perhaps even some of the "successes" claimed by psychotherapists (see Module 25). This of course calls into question the very practice of administering placebos in medical practice, as giving a placebo to a patient is essentially deceitful and dishonest. One could attempt to justify the deceit through a measurable different in the health of a patient or even a perceived difference. Yet for many ethicists, deceit of any sort ought not to be a part of health care, properly administered.

conclusion to draw might be something like "People who eat 200 unwashed Michigan apples each day run a great risk of developing cancer", and that is of limited interest.

Third, an important drawback of many laboratory studies is that they cannot, for ethical reasons, be done with humans. Thus, laboratory rats—which are relatively cheap, numerous, easily obtained, and, most importantly, genetically similar to humans—are used. Genetic similarities notwithstanding, when using laboratory rats there is always the risk, due to certain relevant dissimilarities between rats and humans, that the results obtained may not be transferable to humans. Therefore, conclusions that are generalizable to populations of rats can only be transferred to humans via analogy, and analogical arguments should always be taken with some degree of distrust.

In spite of these problems, randomized controlled experiments are generally the best types of experiments, because of the numerous controls they afford experimenters and the randomness of the selection process for subjects.

Prospective Experiment

A second sort of controlled study is forward-looking or *prospective*. In prospective studies, subjects are not chosen randomly, but rather because of their relationship to the causal factor under investigation in the study. Subjects are selected just because they are known to have the suspected causal factor.

> **In a PROSPECTIVE CONTROLLED STUDY of a suspected causal relationship between two variables α and ω (where α is the suspected cause and ω is the suspected effect), subjects with α form G_E and subjects without α form G_C. If there a statistically significant difference is observed between those in G_E that come to have ω and those in G_C that come to have ω, then there is reason to believe α is a cause of ω.**

For any causal hypothesis under investigation, say, "α is a cause of ω", researchers pick out those subjects with causal condition α (G_E), match this group as closely as possible to another group of subjects without causal condition α (G_C), and then observe over time the number of subjects in G_E that come to have ω in contrast to those of C_G that come to have ω. If the difference is significant, we have reason to believe that α is in fact causally linked to ω.

For example, in one study of 600 diseased patients (from gastrointestinal problems to cancer) out of Bowling Green State University, those who thought God was punishing or abandoning them were 30 percent more likely to die within two years than those who did not think thus.[2] Here the population is "people with diseases", G_C is perhaps "those diseased, who believe in a maleficent deity" and C_E is, presumably, "those diseased, who believe in a beneficent or non-malevolent deity". The conclusion suggested here is that it is not merely belief in a god that translates into longer mortality, but not believing this god is maleficent.

Retrospective Experiment

A third sort of controlled study is backward-looking or *retrospective*. Like prospective studies, in retrospective studies, there is no random selection of subjects. Here subjects that are known to have a particular effect under investigation have their history examined to see if this effect is linked to a suspected causal factor.

> **In a RETROSPECTIVE CONTROLLED STUDY of a suspected causal relationship between two variables α and ω (where α is the suspected cause and ω is the suspected effect), subjects with ω form G_E and subjects without ω form G_C. If there a statistically significant difference is observed between those in G_E that also have α and those in G_C that have α, then there is reason to believe α is a cause of ω.**

For any hypothesis under investigation, say, "α is a cause of ω", researchers pick out those subjects with the suspected effect ω (G_E), match this group as closely as possible to another group of subjects without ω (G_C), and then observe the number of subjects in G_E that have the suspected cause α in contrast to those of C_G that have α. If the difference is significant, we have reason to believe that α is in fact causally linked to ω.

To illustrate, let us take two variables that we know are causally linked: alcohol consumption and memory loss. One way to show that the two are linked might be to do a prospective study—to compare, over time, people who have more than three drinks each day with those who do not. If the experimental group—after a certain number of years, say 20—has considerable difficulty in memory-related tasks as compared to the control group, this gives us some reason to believe that alcohol consumption is causally related to memory loss. As one can imagine, such an approach is less-than-economical, however. It would take considerable time and money. A more efficient approach would be to begin research with a retrospective study. One could compare a small group of people with good memory (G_C), aged 50 or older, to another small group of people, aged 50 or older, with demonstrated memory loss (G_E). If many of the subjects in G_E are heavy drinkers and few in G_C are, there is reason to believe that alcohol consumption is causally related to memory loss.

Women in Science

In the early 20th century, Marie Curie won Nobel Prizes for her work in Physics and Chemistry. In 1983, Barbara McClintock won the Nobel Prize for her distinguished work in Physiology and Medicine. Astronomer and Physicist Jocelyn Bell Burnell discovered pulsars, when she was still a graduate student. Elizabeth Kubler Ross changed the way we think about dying. The list goes on.

Historically, due to ignorance and overt discrimination, women have not played a large part in the development and advance of science. Increasingly today, things are changing; women are playing a greater role. Still the question remains: why are there not more women in science today?

One answer is that the question is ill-formed. Historian of science Naomi Oreskes says, "The question is not why there haven't been more women in science; the question is rather why we have not heard more about them". In other words, the fact that we hear so little of women in science in our texts on science is part of the problem. Perhaps the largest barrier is that women are still the primary caregivers for their children—especially in the United States. Writes Stephen Brush in *American Scientist:*

> By the time a woman lands an assistant professorship, she is likely to be in her late twenties or early thirties. She then has five or six years to turn out enough first-rate publications to gain tenure. If she has children, she must fulfill her family obligations while competing against other scientist who work at least 60 hours a week. If she postpones childbearing, the biological clock will run out about the same time as the tenure clock.

Data-Base Research

Finally, I wish to draw attention to a relatively recent and very trustworthy technique for gleaning causally relevant information about variables. With advancements in computer technology, researchers now often have access to large computer bases of data (e.g., MEDLINE) from which they can collect information for scientific purposes in areas like epidemiological or nutritional studies. Such data-bases are invaluable for gathering causally rich information, where causal links between variables are suspected to exist. Scientific researchers may, for instance, scan such data-bases from effects to causes, like retrospective studies, or from causes to effects, like prospective studies.

What are the key benefits of data-base research? First, as such researchers often work with subjects numbering in the tens of thousands, randomization and controls are not key issues. The large number of subjects, with a small margin of error, makes it a good bet that the data will be representative of the targeted population. Differences that could lead to biases will be swamped out and that makes concern over controls a non-issue. Second, large numbers of subjects make it easy to recognize a causal relationship between variables—or lack of one—without the need for additional research. For instance, if carrots are thought to fight cancer effectively (e.g., as a rich source of β-carotene), then computer-base data of thousands of people, where consumption of carrots is part of the daily diet, should show a marked reduction in the overall incidence of cancer as compared to others, who do not eat carrots at all. Failure to find such a difference might then be a sufficient reason to abandon further research along these lines. Last, data-base research can be time-efficient and cost-effective in many cases. Prospective research, for instance, has the disadvantage that researchers must patiently wait for results—often for decades. Laboratory experiments can sometimes be very expensive, compared to data-base methods.

In all, it is safe to say that the benefits of data-base research methods are incontestable. Computers are making scientific research much more efficient. Moreover, we may add that the results of such research much less ambiguous.

Key Terms

controlled experiment
prospective experimental design
blind experiment
placebo effect
randomized experimental design
retrospective experimental design
double-blind experiment
data-based research

Text Questions

* You want to begin a study that is representative of all people in a particular town, say, Oconomowoc, Wisconsin. What is a good way to select a sample of 100 people that is random?
* Think up a prospective study of your own to examine two causal variables. Once the parameters are set, what are the limits of this experimental design? What are its merits?
* Retrospective studies are generally considered the least reliable of the three controlled studies. Think up a retrospective study of two causal variables. Once the parameters are set, what are the limits and merits of this experimental design? Contrast these results with those from your prospective study.
* How is data-based research preferable to controlled laboratory studies?

Text-Box Questions

* What is the placebo-effect? Why is it so important not only to "blind" subjects but also to "blind" researchers because of it?

* Who is the most famous female scientist that you know? Write a biographical sketch of her.

---

1 Melissa C. Stöppler, "Acute Stress May Boost Immune System", http://stress.about.com/cs/medicalconditions/a/aa111301.htm.
2 Claudia Kalb, "Faith & Healing", *Newsweek,* Nov. 10, 2003, 53.

Module 23
Evaluating Causal Hypotheses

"You ask Carneades, do you, why these things so happen, or by what rules they may be understood? …'Mere accidents', you say. Now, really, is that so? Can anything be an 'accident' which bears upon itself every mark of truth? Four dice are cast and a Venus throw results—that is chance; but do you think it would be chance, too, if in one hundred casts you made 100 Venus throws? It is possible for paints flung at random on a canvas to form the outlines of a face; but do you imagine that an accidental scattering of pigments could produce the beautiful portrait of Venice of Cos?" Cicero, *On Divination*

HAVING EXAMINED BOTH THEORETICAL MODELS AND STATISTICAL HYPOTHESES and having been introduced philosophically to the sticky issue of causation as it relates to the practice of science, we now can turn to an evaluation of causal claims themselves.

Additional Terminology

Although we analyze causal models and claims much the same as statistical models and claims, a causal relationship between values is radically different from one that is statistical. Causal claims are directional, not symmetrical, as causes necessarily precede and bring about their effects.

Let us look more closely at this last statement by means of the following causal hypothesis: "HGH (Human Growth Hormone) causes a significant build up of muscle mass in those who train with weights". What that is saying is that anyone (or just about anyone) who takes HGH and trains with weights will experience a significant gain in muscle mass. Of course, if one could somehow administer HGH to everyone, train them with weights, and watch for gain in muscle mass, one would certainly not see the effect in *every* person. That is because of the variability in individuals' bodily constitutions as well as the overall complex of factors, external as well as internal, that could prevent or even enhance the effect. So what the claim "HGH causes a significant build up of muscle mass in those who train with weights" (or any other causal claim for that matter) really says is "In the main, any person who takes HGH and trains with weights will gain a significant amount of muscle mass", and this of course is not quite a universal claim.

Before turning to the guidelines for evaluation, recall that a causal relationship between values can be positive or negative. Thus, there is need of additional terminology.

Any value V_C is a POSITIVE CAUSAL FACTOR for another value V_E, if V_C produces V_E, and $\sim V_C$ produces $\sim V_E$.

Any value V_C is a NEGATIVE CAUSAL FACTOR for another value V_E, if V_C produces $\sim V_E$, and $\sim V_C$ produces V_E.

Any value V_C is CAUSALLY UNRELATED to another value V_E, if V_C is neither a positive nor a negative causal factor for V_E.

Remember to exercise due caution in your analysis of causal hypotheses and your conclusion. Causal claims need a *substantial* amount of evidence to get them, as it were, off the ground.

Guidelines for Evaluation

Let us now turn to guidelines for evaluating causal models. The principles of evaluation here will be similar to those of correlational models.

Causal Hypothesis: *What is the causal hypothesis under investigation? I.e., state the causal claim as it related to the targeted population.*

Prediction: *What prediction does the causal hypothesis make? State this in such a way that it straightforwardly suggests the experiment/observations to be done.*

Background Information/Research Methods: *Is there background information that is relevant? Is there anything worth mentioning about the research methods? What sort of causal study (i.e., randomized, prospectus, retrospective, or data-base) is being performed? How was the sample selected?*

Results: *What are the relevant data? Be sure to include the sample size and the margin of error thereby indicated. Draw up any diagrams that may be of help in assessing the data.*

Analysis: *If the sample was not random, are there any biases that might make it misrepresentative of the population?*

Is it possible to state precisely the intended population?

What of the sample's size? Is it large enough to offer a sufficiently good representation of the population? If small, is this a pilot study?

Are the data consistent with the causal prediction? If so, to what extent do they confirm it? Are there other hypotheses consistent with the data that cannot be ruled out? If not, must the model be rejected or is there some non-ad-hoc way to salvage it?

Conclusion: *Give you verdict here succinctly and precisely. Be careful here. Causal claims are much stronger than correlational claims, so exercise due caution.*

In the examples that follow, there are data sufficient to set up diagrams and work with the appropriate ME for the subject pools. Yet as some of the exercises in Module 24 show, not all causal episodes will enable you to do this. Do as thorough of a job as you can with the data.

Cigarette Smoking: Reduction & Risk[1]

Smoking and Coronary Heart Disease

A new systematic review of studies on heart disease tells us that smokers with coronary heart disease have have a much greater chance of dying over five years than those with coronary heart disease who quit.

The review examined studies with patients, diagnosed with previous heart attack or stable or unstable angina, who were smokers. The search strategy was extensive, examining nine electronic databases, and the studies were not restricted by language. In all, there were 12,600 patients from 20 studies, mostly using data collected in the 1960s and 1970s. Most cases were men (80%), and average cessation rate of smoking was 45%.

After diagnosis, there were fewer deaths over five years in quitters (1044/5649, 18%) than in people who continued to smoke (1884/6944, 27%), and the degree of reduction was consistent across all death rates reported.

This review gives the consistent and unequivocal answer that smoking remains harmful after a coronary event. Continued smoking after diagnosis of heart disease is causally linked with mortality. Cessation of smoking is no guarantee of a longer life, but it is certainly a more sensible option than continued smoking.

Cigarettes & Cancer: A Brief History of the Debate

1964: "Smoking and Health: Report of the Advisory Committee to the Surgeon General", the first major report in the U. S. to link smoking to lung cancer (in men), is published. The American Medical Association declares that smoking is a health risk and the first national anti-smoking coalition is formed.

1965: Congress passes legislation requiring the following label be placed on every cigarette package: "Caution: Cigarette Smoking May be Hazardous to Your Health".

1970: The Public Health Cigarette Smoking Act is passed and prohibits advertising on radio and television. It also requires a changed label on cigarette packages: "Warning: The Surgeon General Has Determined that Cigarette Smoking is Dangerous to Your Health".

1973: Smoking is banned on commercial airline flights that are under six hours.

1982: Congress temporarily doubles the federal excise tax on cigarettes (now $0.16 per pack); the last increase was in 1951.

1984: The Surgeon General puts forth the goal of a "smoke-free society" by 2000.

1986: A report of the Surgeon General focuses on the risks of second-hand smoke. The "temporary" tax is made permanent and an excise tax is placed on smokeless tobacco.

1991: The federal excise tax on a package of cigarettes jumps to $0.20. The FDA approves a nicotine patch as a prescription drug.

1993: The federal excise tax increases to $0.24.

1996: President Clinton announces a major initiative to prevent children and adolescents from acquiring and using tobacco. The Food and Drug Administration plans to regulate the sale and distribution of cigarettes and smokeless tobacco to children and adolescents. It plans to make the major tobacco companies use multimedia to educate young people about the health risks dangers of tobacco.

1997: R. J. Reynolds voluntarily removes "Joe Camel" from all advertisements.

1998: California bans smoking in bars. Documents are disclosed that suggest that R. J. Reynolds was targeting the youth with its Joe-Camel advertisements.

1999: Before congress, tobacco representatives confess that nicotine is addictive and agree that it may be a cause of cancer. Forty six states and five territories win a 206 billion dollar settlement to settle Medicaid lawsuits. Philip Morris grudgingly acknowledges the wealth of data that supports a causal connection between smoking and cancer. "There is an overwhelming medical and scientific consensus that cigarette smoking causes lung cancer, heart disease, emphysema and other serious diseases in smokers".

2000: Federal excise tax increases to $0.34 per package.

2001: President Clinton issues an Executive Order stating that the United States will be a leader on global issues relating to tobacco.

2002: Federal excise tax per package increases to $0.39.

Present: There is current global and national debate about enlarging the warnings on cigarette packages or using pictures to deter smokers. The federal excise tax is still at $0.39. State taxes vary from $0.05 and $0.07 for North and South Carolina to $2.20 and $2.46 for New Jersey and Rhode Island! As of January 1, 2005, the average state tax is $0.79. As of Title IX, Chapter 36, Section 1333, each pack of cigarettes sold in the U.S. must have one of the following warnings: Smoking Causes Lung Cancer, Heart Disease, Emphysema, and May Complicate Pregnancy; Quitting Smoking Now Greatly Reduces Serious Risks to Your Health; Smoking By Pregnant Women May Result in Fetal Injury, Premature Birth, and Low Birth Weight; or Cigarette Smoke Contains Carbon Monoxide.

***Assessment of* Smoking and Coronary Heart Disease**

Causal Hypothesis

For those diagnosed with heart disease, continued smoking is a *positive causal factor* for mortality.

Prediction

If the causal hypothesis is true, then *those diagnosed with heart disease, who continue to smoke, should have a greater mortality rate that those who do not.*

Background Information/Research Methods

* The research is data-base and prospective in design.
* Nine electronic databases, not restricted by language, were used.
* In all, there were 12,600 patients from 20 studies, mostly using data collected in the 1960s and 1970s. Most cases were men (80%), and average cessation rate of smoking was 45%.

Results

* 20 studies with about 12,600 patients, mostly from data collected in the 1960s and 1970s: 5649 in the control group (G_C = 5649), those with heart disease who quit smoking; 6944 in the experimental group (G_E = 6944), those with heart disease who still smoke.
* Most cases were men (80%), and average cessation rate was 45%.
* There were fewer deaths in quitters (1044/5649, 18%) than in those who continued to smoke (1884/6944, 27%).
* The strength of causal relationship ranges from .11 (.28 - .17) to .07 (.26 - .19).

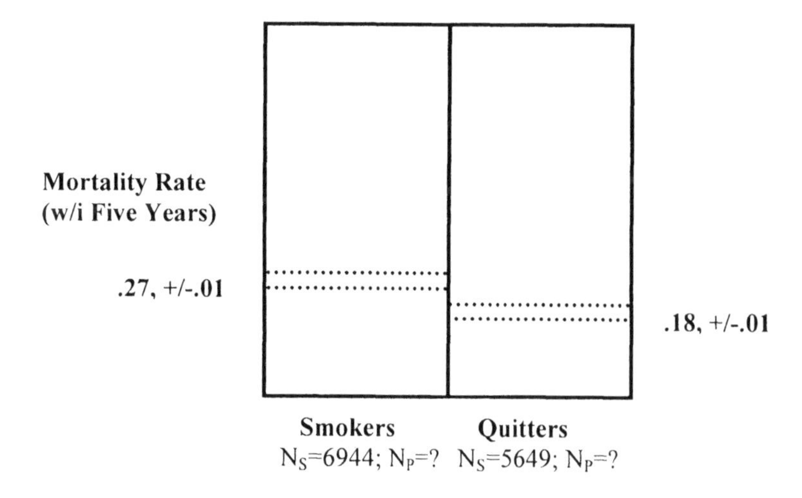

Figure 23.1 Mortality Rate: Smokers vs. Quitters. Comparison of the mortality rate of CHD patients who continued to smoke after diagnosis with those who quit.

Analysis

As this was a data-base study, subjects could not be chosen randomly. With close to 13,000 subjects examined, there is little reason to cast any doubt upon the findings. The data are consistent with the hypothesis and offer substantial confirmation of it. We are in a relatively good position to suspect a causal link between the two variables here.

Conclusion

For those diagnosed with heart disease, smoking is indeed a *positive causal factor* for mortality.

Benacol and Lipids[2]

Can Benecol Reduce Cholesterol?

A study in European Journal of Clinical Nutrition *analyzed the effect of Benecol in reducing blood cholesterol. J.A. Westrate and G.W. Meijer gave five groups of subjects 30 grams per day of a specific type of margarine. They then examined the blood cholesterol levels over te next three-and-one-half weeks. Blood samples were taken at the two-and-one-half week point as well as at the end of the study, and the average values of these two measurements were taken. The study was randomized, double-blind, and cross-over and seemed to be conducted with great detail.*

One hundred adults (50 men and 50 women) were randomly selected for this study. On average, subjects had a mean body mass of 24 kilograms per meter and a mean age of 45 years (from an allowable range 18 to 65 years). Mean values for lipids were:

Total cholesterol 5.4 +/- 1.1 mmol/L
LDL cholesterol 3.5 +/- 1.0 mmol/L
HDL cholesterol 1.3 +/- 0.35 mmol/L

For total cholesterol, mean reductions were 0.37 mmol/L for Benecol and 0.43 mmol/L for the soybean-sterol enriched spread. For LDL cholesterol, mean reductions were 0.40 mmol/L for Benecol and 0.44 mmol/L for the soybean-sterol enriched spread. There was no noticeable difference in men and women. The same degree of reduction was found regardless of the differences in starting total and LDL cholesterol of subjects. Margarines enriched with the three other plant oils had no noticeable effect on total and LDL cholesterol.

This study confirmed other studies that demonstrate the quick fall in LDL and total cholesterol using the types of plant sterols found in Benecol and soybean-based spreads. The study indicates that using Benecol or a soybean-based spread could have a large impact over time in reduction and regulation of blood-cholesterol levels.

***Assessment of* Can Benecol Reduce Cholesterol?**

Causal Hypothesis

Benecol is a *negative causal factor* for total cholesterol and LDL cholesterol.
Prediction

If the causal hypothesis is true, then subjects ingesting 30 grams of Benecol each day should show a marked reduction in total cholesterol and LDL cholesterol as opposed to those who use other margarines.

Background Information/Research Methods

The experiment was a randomized, double-blind crossover study that was conducted in three and one-half weeks. It was detailed in its measurement of diets, energy intake, and blood chemistry.

Results

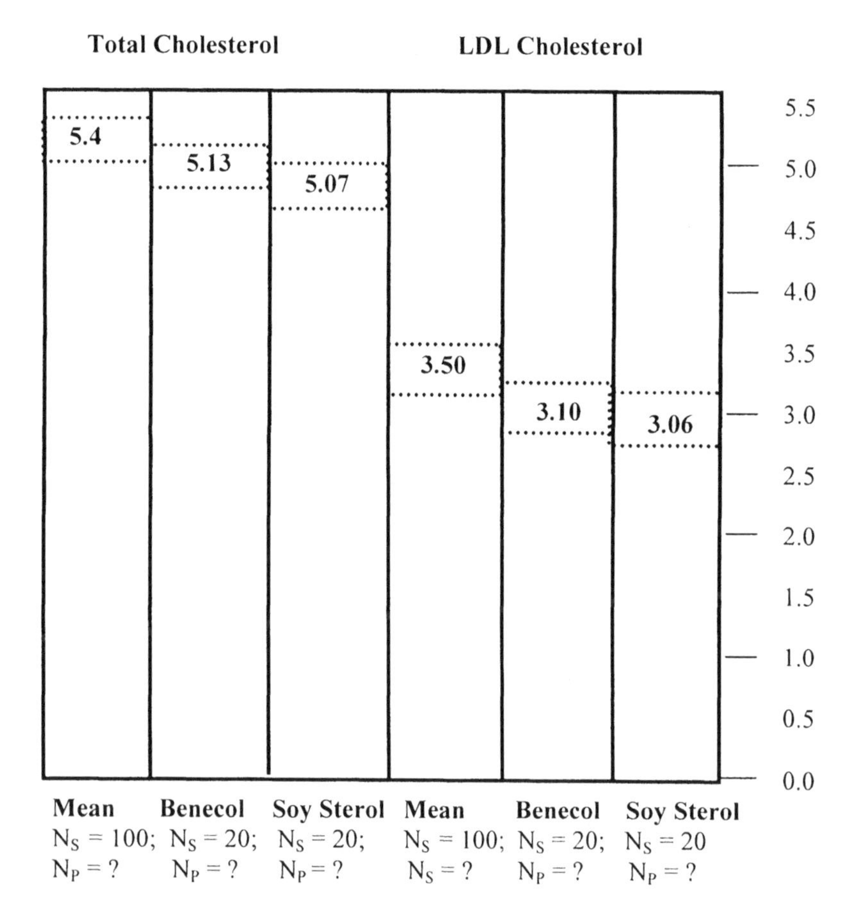

Figure 23.2 Mean blood cholesterol (total and LDL) of all subjects as compared to the blood cholesterol (total and LDL) of those in Benecol and soy-sterol groups (ME = +/-.22).

* 50 men and 50 women, with a mean body mass of 24 kg/sq meter (allowable range 19 to 30 kilograms per square meter) and a mean age of 45 years (allowable range 18 to 65 years), were chosen.
* The mean values for lipids were:
  Total cholesterol 5.4 ± 1.1 (SD) mmol/L
  LDL cholesterol 3.5 ± 1.0 mmol/L
  HDL cholesterol 1.3 ± 0.35 mmol/L
* Subjects were sorted into groups and given five margarines (30 grams per day): Flora, Benecol, and spreads enriched with soybean, ricebran and sheanut oils. (No is information given about the numbers in these groups, but we can assume that there are 20 per group and that the number of men and women per group were roughly the same.)
* Blood samples were taken 2.5 and 3.5 weeks after starting on a margarine, and the average values of these two measurements were taken.
* Mean reductions in total cholesterol:
  Benecol: 0.37 mmol/L
  Soybean Sterol: 0.43 mmol/L
* Mean reduction in LDL cholesterol:
  Benecol: 0.40 mmol/L
  Soybean Sterol: 0.44 mmol/L
* Benecol and the spread enriched with esterified soybean sterols significantly reduced total cholesterol and LDL cholesterol by about 8% and 13% respectively in both men and women.
* The margarines enriched with other plant oils were without effect. HDL cholesterol was unchanged.
* All the data are not given, so we can only diagram what we are given. (I use a little ingenuity in Figure 23.2, below.) I depict the reduction of total cholesterol and LDL cholesterol as they relate to mean total cholesterol and mean LDL cholesterol.

Analysis

The report said the study confirmed other studies demonstrating rapid falls in LDL and total cholesterol using spreads with Benecol and soy sterols. With the relatively few number of subjects in the Benecol and soy-sterol groups, we are certainly not in a position to state unequivocally that Benecal (and soy-sterol margarines) are both a negative causal factor for total and LDL cholesterols. The results, however, are promising.

Conclusion

At this point, we can say that it is quite possible that Benecol is a negative causal factor for total and LDL cholesterol. More research needs to be conducted.

Key Terms

positive causal factor	negative causal factor
causally unrelated	rationalist Greek medicine

empiricist Greek medicine

EXERCISE

The following was found on a box of oatmeal from a major manufacturer.

Great News! Town Confirms Oatmeal May Help Lower Cholesterol

100 people in Lafayette, Colorado volunteered to eat a good-sized bowl of oatmeal for 30 days to see if simple lifestyle changes—like eating oatmeal—could help reduce cholesterol. After 30 days, 98 lowered their cholesterol. With these great results, the people in Lafayette proved to themselves that simple changes can make a real difference.

Without challenging the data, can you think of reasons to be suspicious of what the study here suggests?

TEXT QUESTIONS

* In what key ways do causal hypotheses differ from statistical hypotheses?
* Using the study on smoking and coronary heart disease as your guide, how does data-base research make causal analysis much more efficient?

TEXT-BOX QUESTIONS

* Using the text-box on cigarettes and cancer as a guide, why are causal claims so difficult to substantiate?
* What was the debate between rationalists and empiricists in early Greek medicine? How did causality come into play?

---

1 J. A. Critchley and S. Capewell, "Mortality Risk Reduction Associated with Smoking Cessation in Patients with Coronary Heart Disease: A systematic review", *JAMA* 2003, 290: 86-97.

2 J. A. Westrate & G. W. Meijer, "Plant Sterol-Enriched Margarines and Reduction of Plasma Total- and LDL-Cholesterol Concentrations in Normocholesterolaemic and Mildly Hypercholesterolaemic Subjects", *European Journal of Clinical Nutrition*, 1998 52: 334-43.

Module 24
Exercises for Causal Hypotheses

"This relativity of causes to interests, and to background condiitons not mentioned in the 'hard science' explanation of the event in question, does not make causation something we simply legislate.... Our conceptual scheme restricts the 'space' of descriptions to us; but it does nto predetermine the answers to our questions". Hilary Putnam, *The Many Faces of Realism*

EVALUATE EACH OF THE FOLLOWING EPISODES according to the guidelines summed below.

CAUSAL HYPOTHESIS: *What is the causal hypothesis under investigation? I.e., state the causal claim as it related to the targeted population.*
PREDICTION: *What prediction does the causal hypothesis make? State this in such a way that it straightforwardly suggests the experiment/observations to be done.*
BACKGROUND INFORMATION/RESEARCH METHODS: *Is there background information that is relevant? Is there anything worth mentioning about the research methods? What sort of causal study (i.e., randomized, prospectus, retrospective, or data-base) is being performed? How was the sample selected?*
RESULTS: *What are the relevant data? Be sure to include the sample size and the margin of error thereby indicated. Draw up any diagrams that may be of help in assessing the data.*
ANALYSIS: *If the sample was not random, are there any biases that might make it misrepresentative of the population?*
Is it possible to state precisely the intended population?
What of the sample's size? Is it large enough to offer a sufficiently good representation of the population? If small, is this a pilot study?
Are the data consistent with the causal prediction? If so, to what extent do they confirm it? Are there other hypotheses consistent with the data that cannot be ruled out? If not, must the model be rejected or is there some non-ad-hoc way to salvage it?
CONCLUSION: *Give you verdict here succinctly and precisely. Be careful here. Causal claims are much stronger than correlational claims, so exercise due caution.*

Green Tea & Cancer[1]

Can Consumption of Green Tea Prevent Gastric Cancer?

There have been some studies that suggest that consumption of green tea reduces the risk of gastric cancer. One study in 2001, with a large number of Japanese people, aimed to shed further light on this suspected causal link.

Researchers examined 26,311 Japanese (11,902 men and 14,409 women), each of whom was at least 40 years old. They were selected from three municipalities of a region in northern Japan, which is known for its high incidence of gastric cancer. Beginning in 1984, subjects completed questionnaires on many aspects related to their health (dietary intake, smoking and drinking habits, family history of disease, etc.). They were also asked how much green tea they drink on a daily basis. These subjects were then watched for nine years.

In accordance with their daily consumption of green tea, subjects were divided into four categories: less than one cup a day (19% of all subjects); one or two cups a day (17%); three or four cups a day (22%); and five or more cups a day (42%) (a typical cup contains 100 mls. of tea).

From the local cancer registery, a total of 419 cases of gastric cancer were discovered among these subjects (296 in men and 123 in women) over time. Results were adjusted for gender, age, history of peptic ulcer, smoking status, alcohol intake, daily consumption of rice, black tea, coffee, meat, green and yellow vegetables, pickled vegetables, fruit, bean-paste soup, and socio-economic status. It was discovered that there was no link between green-tea consumption and the risk of gastric cancer among the subjects and there were no differences across genders.

Cups of Green Tea/Day	**-1 (19%)**	**1-2 (17%)**	**3-4 (22%)**	**+4 (42%)**
Cases of Gastric Cancer	**66**	**68**	**79**	**206**

Figure 24.1. Green Tea and Cancer. Number of cases of gastric cancer related to consumption of green tea.

This was a prospective study with a large number of subjects (over 26,000) that is free of the biases of earlier retrospective (case-controlled) studies that have suggested a link between green tea and gastric cancer. Notwithstanding these methodological advantages over prior research, this study suffers from a small methodological defect: Results could not be adjusted for participants' history of cancer. When the survey was being done, most cancer patients were not informed of their status and so information about their prior history of cancer was not available.

Obesity vs. Smoking[2]

Causes of Death in the United States

Americans' eating habits are becoming so bad lately that poor diet and physical inactivity may soon overtake tobacco as the leading cause of death. This is the conclusion of researchers led by Ali Mokdad at the Centers of Disease Control and Prevention in Atlanta. What is worse, smoking, poor diet, and inactivity are preventable behaviors.

Concerning the study's design, the authors write:

> *Comprehensive* MEDLINE *search of English-language articles that identified epidemiological, clinical, and laboratory studies linking risk behaviors and mortality. The search was initially restricted to articles published during or after 1990, but we later included relevant articles published in 1980 to December 31, 2002. Prevalence and relative risk were identified during the literature search. We used 2000 mortality data reported to the Centers for Disease Control and Prevention to identify the causes and number of deaths. The estimates of cause of death were computed by multiplying estimates of the cause-*

attributable fraction of preventable deaths with the total mortality data.

Fast Food or Slow Living?

McDonald's Corporation refuses to acknowledge that it may be partly at fault for the rise in child obesity. At an investor's conference (Banc of America Securities) on Sept. 20 of 2006, Chief Financial Officer Matthew Paull said that the main cause is American's inactive lifestyle.. "If you eat too much of anything, it's a problem, if you don't exercise". He went on to say that McDonald's is not the problem. "There's no such thing as good food and bad food". Paull noted that its average customer in the U.S. eats three times a month at McDonald's and takes 87 meals "someplace else". McDonald's or any other restaurant should not be blamed for peoples' habits. "We do not believe that any restaurant is the cause of the problem".

Paull did state that child obesity was a concern of McDonald's and that the restaurant chain wanted to be part of the solution. "We're big. We're powerful. We can make a difference". Chief Executive Jim Skinner added that McDonald's would advertise "Happy Meal combos that fit into the recommended daily allowance guidelines", listed by the government. Neither Paull nor Skinner elaborated on this.

Paull anticipated and deflected criticism that McDonald's was motivated by existing or potential lawsuits. "It's not about lawsuits. It's about being responsible". He illustrated by McDonald's introduction of a chicken-based snack wrap in August. The chicken-wrap was designed chiefly to increase afternoon sales, but it also shows their commitment to a healthier lifestyle.

This study yielded the following results. Of all deaths in the United States, the leading causes were:

tobacco (435,000 deaths; 18.1% of all U. S. deaths),
poor diet and physical inactivity (400,000 deaths; 16.6%), and
and alcohol consumption (85,000 deaths; 3.5%).

Other identifiable causes of death were:

microbial agents (75,000),
toxic agents (55,000), motor vehicle crashes (43,000),
incidents involving firearms (29,000),
sexual behaviors (20,000),
and illicit use of drugs (17,000).

A 1990 study by Michael McGinnis and William Foege shows 400,000 deaths due to tobacco as compared to 300,000 deaths due to obesity. Obesity is quickly catching up to tobacco as the leading cause of death.

"There's been a big narrowing of the gap", Mokdad, told the Washington Post. *"Physical inactivity and poor diet is still on the rise. So the mortality will still go up. That's the alarming part—the behaviour is still going in the wrong direction".*

If current trends continue, obesity will become the leading cause by 2005, with the toll surpassing 500,000 deaths annually. "This is a tragedy", says CDC director Julie Gerberding. "We are looking at this as a wake-up call".

While the gap between deaths due to poor diet and inactivity and those due to smoking has narrowed substantially, this is not due to a decrease in smoking deaths. "The most disappointing finding may be the slow progress in reducing tobacco-related mortality", the team notes. In re-

sponse to the study's stark findings, the U.S. Department of Health and Human Services launched a national education campaign and research strategy on Tuesday".

"Americans need to understand that overweight and obesity are literally killing us", said health secretary Tommy Thompson, adding that the study's findings "should motivate all Americans to take action to protect their health".

The Atkins' Diet & Long-Term Weight Loss[3]

Is the Atkins' Diet a "Hopeless Fad"?

Two studies in a prestigious medical journal have caused a media maelstrom. They show that the Atkins regimen can shed weight quickly...but the pounds are soon regained.

Both studies appeared in The New England Journal of Medicine, *with each randomly dividing subjects into two groups. In each study, one group was told to reduce caloric intake while the other wasn't expressly told to cut calories, but rather instructed to keep carbohydrate intake extremely low as dictated by the late Dr. Robert Atkins' books. One, conducted at the Philadelphia Veterans Administration Hospital, lasted six months, comprised subjects with an initial average weight of about 215 pounds. The other was conducted at three different centers, lasted 12 months, and comprised subjects with an initial average weight of about 290 pounds.*

The six-month study found that Atkins dieters lost weight at about twice the rate as the higher-carb group—for two months. Thereafter neither group lost much weight. By the end of six months, the Atkins dieters, however, had still managed to keep off about twice as much weight as the higher-carb group. The average loss was a mere 13 pounds from that original 290.

In the 12-month study, the Atkins group lost considerably more weight for the first half year, but began packing the pounds back on, and did so faster than the higher-carb group. Ultimately, concluded the researchers, "the differences were not significant at the end one of year".

The probable explanation for the early weight loss, said the chief researcher of the 12-month study (Gary Foster of the University of Pennsylvania), is that it "gives people a framework to eat fewer calories, since most of the choices in this culture are carbohydrate driven. You're left eating a lot of fat, and you get tired of that". Consequently, the Atkins plan is merely a low-calorie diet in disguise.

Soon though, Atkins dieters become so starved for carbohydrates that they either start cheating or quit the plan altogether. In fact, both studies were plagued by high dropout rates from all sets of dieters. The only weight-control regimens that work for life require both eating in moderation and exercise.

Foster's co-researcher James Hill of the director of the University of Colorado Center for Human Nutrition in Denver noted that we must be cautious. All diets that focus on calorie-cutting work. "The Atkins diet produces weight loss, as does the grapefruit diet, the rotation diet, and every other fad diet out there". He added, "I haven't seen any data anywhere saying Atkins is better than these other diets for weight loss".

The New England Journal of Medicine's *findings also belied the assertions in a massively publicized* New York Times Magazine *cover story last year by Gary Taubes that rocketed Atkins' diet-book sales into the stratosphere. It also landed Taubes his own $700,000 book-contract. In the article, Taubes ignored red flags in the Atkins book that flapped as if in a hurri-*

cane. In addition to its something-for-nothing weight-loss promise, Atkins also insisted his diet relieves "fatigue, irritability, depression, trouble concentrating, headaches, insomnia, dizziness, joint and muscle aches, heartburn, colitis, premenstrual syndrome, and water retention and bloating"—in short, it does everything but walk the dog.

"Worse, Taubes ignored a mass of published and peer-reviewed studies showing low-carbohydrate diets to be ineffective for long-term weight loss, such as a review in the April 2001 Journal of the American Dietetic Association *of "all studies identified" that looked at diet nutrient composition and weight loss. It claimed to have found over 200, with "no studies of the health and nutrition effects of popular diets in the published literature" excluded. In some, subjects were put on "ad libitum" diets—meaning they were allowed to eat as much as they wanted as long as they consumed fat, protein, and carbohydrates in the directed proportions. In others, subjects were put on controlled-calorie diets that also had directed nutrient proportions. The conclusion is this: Those who ate the least fat carried the least fat.*

Instead, Taubes put his own spin on five then-unpublished research efforts. "The results of all five of these studies are remarkably consistent. Subjects on some form of the Atkins diet", he insisted, "lost twice the weight as the subjects on the low-fat, low-calorie diets".

In the article, Taubes cited five then-unpublished studies claiming: "The results of all five of these studies are remarkably consistent. Subjects on some form of the Atkins diet", he insisted, "lost twice the weight as the subjects on the low-fat, low-calorie diets". Two of those, however, are the ones discussed here, where differences at the end of one year were not significant. Somehow, "The differences were not significant at the end of one year" doesn't seem to support "lost twice the weight". Three of the five studies lasted only 12 weeks, which these two studies indicate to be obviously of no use. A fourth, according to chief author Eric Westman at Duke University, does back Atkins, but then again, Atkins backs him.

The Atkins Center has an open-ended commitment to support fully Westman's work. The last of the five studies is from the University of Cincinnati, where co-author Randy Seeley of the university's Obesity Research Center says the Atkins cohort did have twice the weight loss at the end of the six months. Nevertheless his explanation is similar to and perhaps more colorful than that of Foster. "If you're only allowed to shop in two aisles of the grocery store, does it matter which two they are?" he asks. However, at least there seemed no evidence that all that saturated fat in the Atkins diet increases the risk of heart disease. In neither New England Journal of Medicine *study did the Atkins dieters have increased LDL (low-density lipoprotein or "bad cholesterol"), and the 12-month study even found a small increase in HDL high-density lipoprotein or "good cholesterol". Last, triglycerides (fatty compounds in blood) of the Atkins dieters decreased. Lower triglyceride levels have been linked to lower rates of heart disease.*

One should be cautious of such findings, says Robert H. Eckel, professor of medicine at the University of Colorado Health Sciences Center, where Hill works. Eckel, who coauthored an accompanying commentary in the New England Journal of Medicine, *says one probable reason for improved blood readings in the Atkins cohorts is that they did have greater weight loss, at least part of the time, in both studies. "Generally when people lose weight, both triglycerides and HDL improve", he stated. To be sure, even the higher-carbohydrate losers showed some improvement in both measurements. As to HDL, he says, not all HDL is created equal. Just as we once thought all cholesterol was bad, there is now evidence that some "good" HDL may not be good after all. Finally, says Eckel, epidemiological studies indicate that triglycerides appear to have only a mild direct impact on heart disease; rather they are a marker for other factors that do impact it, just as open umbrellas are not the cause of rain but markers thereof.*

The Atkins Center was overjoyed that the new studies may indicate the regimen is not dangerous. How peculiar when the most you can say for the best-selling fad-diet book of all time is that it probably does not kill people.

Irresponsible Medical Practice?

No other diet has caused such controversy as the Atkins diet. Why? It seems to deliver on its promise for quick weight loss and do so in a way that defies conventional dietary wisdom. It allows adherents to eat many of those foods that they've all along been told to stay away from—bacon, eggs, meats, and cheese—and still lose weight. Studies, to date, seem to confirm the effectiveness of the diet for weight loss, but questions remain. First, is this diet any more effective than others for weight loss? Second, are there unknown, long-term health risks that might make the diet a poor choice?

A statement from the American Heart Association advises due caution before considering any high-protein diet:

> High-protein diets have recently been proposed as a "new" strategy for successful weight loss. However, variations of these diets have been popular since the 1960s. High-protein diets typically offer wide latitude in protein food choices, are restrictive in other food choices (mainly carbohydrates), and provide structured eating plans. They also often promote misconceptions about carbohydrates, insulin resistance, ketosis, and fat burning as mechanisms of action for weight loss. Although these diets may not be harmful for most healthy people for a short period of time, there are no long-term scientific studies to support their overall efficacy and safety. These diets are generally associated with higher intakes of total fat, saturated fat, and cholesterol because the protein is provided mainly by animal sources. In high-protein diets, weight loss is initially high due to fluid loss related to reduced carbohydrate intake, overall caloric restriction, and ketosis-induced appetite suppression. Beneficial effects on blood lipids and insulin resistance are due to the weight loss, not to the change in caloric composition. Promoters of high-protein diets promise successful results by encouraging high-protein food choices that are usually restricted in other diets, thus providing initial palatability, an attractive alternative to other weight-reduction diets that have not worked for a variety of reasons for most individuals. High-protein diets are not recommended because they restrict healthful foods that provide essential nutrients and do not provide the variety of foods needed to adequately meet nutritional needs. Individuals who follow these diets are therefore at risk for compromised vitamin and mineral intake, as well as potential cardiac, renal, bone, and liver abnormalities overall.

Atkins himself passed away in April of 2003. The explanation had been that he slipped on a patch of ice and suffered severe head injuries. The story itself has recently been called into question.

Atkins' health records were made public to an anti-Atkins group called Physicians' Committee for Responsible Medicine (see Atkinsdietalert.org), which sent the medical examiner's report to the *Wall Street Journal* in February of 2004. Atkins, it seems, had a history of heart disease (high blood pressure, congestive heart failure, and heart attack). Those in the Atkins camp insist that his heart problems began three years ago, when he suffered a viral infection. Many experts insist that a more plausible explanation is his high-fat diet. The truth, however, may be impossible to know, as no autopsy was conducted; his body was conviently cremated.

The worry of several nutritionists and other health-care experts is that the diet may prove no more effective than all other calorie-restrictive diets in the long term and that it may have heretofore undisclosed health risks that outweigh the short-term benefits of quick weight loss.

Assessment of Can Consumption of Green Tea Prevent Gastric Cancer?

Causal Hypothesis

Prediction

Background Information/Research Methods

Results

Analysis

Conclusion

Assessment of Causes of Death in the United States

Causal Hypothesis

Prediction

Background Information/Research Methods

Results

Analysis

Conclusion

***Assessment of* Is the Atkins' Diet a Hopeless Fad?**

Causal Hypothesis

Prediction

Background Information/Research Methods

Results

Analysis

Conclusion

---

1 Y. Tsubono et al., "Green Tea and the Risk of Gastric Cancer in Japan", *The New England Journal of Medicine,* 2001, 344: 632-636.
2 Ali H. Mokdad, James S. Marks, Donna F. Stroup, Julie L. Gerberding, "Actual Causes of Death in the United States, 2000", *Journal of the American Medical Association,* 2004; 291:1238-1245.
3Michael Fumento, "Hopeless Fad: Sorry the Atkins Diet Still Doesn't Work", http://www.nationalreview.com/comment/comment-fumento060603.asp.

PART IV

SCIENCE & ITS PRETENDERS

SECTION EIGHT

Science & Non-Science

Module 25
Science, Marginal Science & Pseudo-Science

"The year 1999 seven months
From the sky will come the great King of Terror.
To resuscitate the great king of the Mongols.
Before and after Mars reigns by good luck". Nostradamus, *Quatrains* IX.72

THIS MODULE LOOKS AT PRACTICES that are not quite up to the standards of normal science as well as those that claim scientific status, but, upon inspection, have nothing scientific about them.

Criteria of Adequacy

As we have seen in the first half of this book, there is no consensus among scientists or philosophers of science on the nature of science. Some philosophers of science argue that science can be differentiated from marginal science by stepping back from the practice of science and asking the right sort of questions that enable us, as best we can, to differentiate between the two. On this view, the scientist pursues an explanation of the facts, while the philosopher of science is a criteriologist (see text box, next page), whose job is to analyze the procedures and logic of scientific explanation.

Before we begin categorizing certain practices as scientific or unscientific, let us specify criteria that help us in distinguishing the one from the other. These criteria are called the *criteria of adequacy*. They are used to decide among competing hypotheses and to assist us in assessing whether a practice is scientific, marginally scientific, or pseudo-scientific. It is important to realize that these criteria are not foolproof guides to disambiguation.

The first such criterion is testability.

> **TESTABILITY:** *To be scientific, a hypothesis must yield an unambiguously testable prediction, other than what it was introduced to explain.*

To put this another way, there must be some conceivable state of affairs (e.g., an experiment to be performed) that, if true, would show the hypothesis to be false. If a hypothesis is introduced simply because it is consistent with certain data under scrutiny, it is said to be *ad hoc*. On the one hand, as we saw with Mendel's hypothesis on inherited characteristics (see Module 15), sometimes data by themselves, having been gathered and thoroughly examined, seem to point to a particular hypothesis. There is, of course, nothing wrong with forming a hypothesis in *ad hoc* fashion, but consistency with the data is by itself not sufficient to show that one hypothesis is preferable to another. There may be many hypotheses consistent with the data. Thus, hypotheses must have testable predictions so that we can decide between them. On the other hand, genuinely *ad hoc* hypotheses are usually introduced in quite suspicious ways. Within the overall framework of a scientific theory, when certain data are found to be inconsistent with that theory, supporters of the theory may add a hypothesis *ad hoc* just to make the data consistent with it (as was the case with the planet Vulcan in the framework of Newtonian dynamics, see Module 15).

Second, there is the criterion of simplicity.

> **SIMPLICITY:** *Of competing hypotheses, the hypothesis that makes the fewest assumptions about the nature of reality is most likely to be true.*

Going back to our comparison of Ptolemy's geocentric model of the universe and Copernicus' heliocentric model in Module 16, we found Copernicus' explanation of the proximity of the orbits of Mercury and Venus to the sun to be, in certain key respects, much simpler than that of Ptolemy. Ptolemy had to posit that Mercury and Venus move on the same deferent arm as the sun, which is again *ad hoc*. Also, Alcide d'Orbigny's attempt to salvage Catastrophism (Module 6, textbox) by positing 27 special creations by a beneficent deity was a clear case of scientific prodigality, not simplicity.

Science, Non-Science, & Philosophy of Science

Philosopher John Losee argues that philosophers of science are critics of the practice of scientists, whose work allow for a crucial distinction between valid scientific practice and what is marginally scientific or pseudo-scientific. He writes:

> Philosophy of science is a second-order criteriology. The philosopher of science seeks answers to such questions as:
>
> 1. What characteristics distinguish scientific inquiry from other types of investigation?
> 2. What procedures should scientists follow in investigating nature?
> 3. What conditions must be satisfied for a scientific explanation to be correct?
> 4. What is the cognitive status of scientific laws and principles?

To ask these questions is to assume a vantage-point one step removed from the practice of science itself.

Third, there is the criterion of fruitfulness.

> **FRUITFULNESS:** *Of competing hypotheses, that hypothesis is best which makes the most novel predictions.*

Let us illustrate by Freud's use of the concept of repression—the burying of some painful truth (generally, sexual) in the Unconscious. Repression is the cornerstone of Freudian psychoanalysis. Yet Freud sometimes used his theory of repression in a tendentious and spurious way to "confirm" his theory of it. He was wont to say of those who vigorously denied his theory of psychosexual repression that they were themselves highly repressed. Thus, opposition to his theory only proved it to be correct. Clearly, this is a theory that takes few risks as far as predictions go. It is not fruitful, as it is hard to say just what it would take to prove his theory wrong. In contrast, Einstein's view of universal gravitation is very fruitful, as the theory makes many new and bold predictions (very many of which have been put to the test and confirmed, see Modules 1 and 13).

Fourth, there is the criterion of scope.

Scope: *Of competing hypotheses, the hypothesis that has the greatest scope (i.e., predicts and explains more than other hypotheses) is most likely to be true, if its predictions turn out true.*

When comparing Newton's theory of universal gravitation with that of Einstein, the latter's theory has greater scope in that it explains all the phenomena that Newton could explain as well as a few phenomena that are known to be inconsistent with Newton's dynamics (e.g., the anomalous perihelion of Mercury, see Module 11).

Last, there is the criterion of conservatism.

Conservatism: *Of competing hypotheses, the hypothesis that squares best with generally accepted scientific beliefs at a particular time is most likely to be true.*

This certainly makes much sense, on the assumption that science is cumulative and progressive. However, it can be a hindrance to progress. For instance, at the time that Copernicus' theory was published, even though his views were nearer to the truth than those of Ptolemy, heliocentrism directly challenged many of the entrenched views of Aristotelian physics and astronomy. Thus, by the criterion of conservatism alone, Ptolemy's theory was more likely to be true, although we now know it to be false. To get heliocentrism off the ground, there was need of a new physics, which did not exist at the time.

Given the criteria above, we are now in a position to distinguish between hypotheses that are *scientific*, those that are *marginally scientific*, and those that are merely *pseudo-scientific*.

A hypothesis is SCIENTIFIC (1) if it is articulated with regard for simplicity, fruitfulness, scope, and conservatism *and* (2) if it makes at least one prediction that can, at least in principle, be put to an unambiguous test.

A hypothesis is MARGINALLY SCIENTIFIC (1) if it is articulated with regard for simplicity, fruitfulness, scope, and conservatism *and,* (2) though it makes at least one prediction, that prediction cannot, even in principle, be put to an unambiguous test.

A hypothesis is PSEUDO-SCIENTIFIC (1) if it is articulated with little regard for simplicity, fruitfulness, scope, and conservatism *or* (2) if it fails to generate at least one prediction that can, at least in principle, be put to a test.

Note a few things here. First, the difference between a scientific hypothesis and one that is marginally scientific is that the former can be put to an *unambiguous* test (Einstein's relativity theory), while the latter cannot (Freud's theory of repression). This condition, of course, makes testability through at least one unambiguous prediction a necessary condition for a hypothesis to be genuinely scientific. The other four criteria should not be taken as necessary conditions, but rather as desiderata (things desirable, but not needed). Second, the notion of *in-principle testability* is given here to allow for hypotheses that cannot now be tested due to limits of current technologies. For example, 100 years ago there was no way to test for the hypothesis "The dark side

An It-Would-Certainly-Seem-Possible Interpretation of Quatrain X.72

The following interpretation of Quatrain X.72 (beginning of this module) is from on the web page of the Nostradamus Society of America. It speaks for itself!

In the year 1999, in the seventh month (suggesting July, although the French word "sept" might suggest the abbreviation for the English word September), from the sky (suggesting an aerial event, perhaps through means of aircraft, missiles or something else) will come the great (as in powerful) King of Terror (suggesting an unnamed leader of a nation who possess a powerful military, probably a foe of Europe and the West), bringing back to life (rebirthing) the great King of the Mongols (Genghiz Khan was the Mongolian king who conquered Asia in the 13th century, thus an Asian military leader seems to be suggested by this phrase). Before and after (some accident or military event in July of 1999), Mars (the symbolic planet of war in astrology) to reign by good fortune" (war appears to reign fortunate for an unnamed nation, or group of nations, "before and after" July of 1999).

If "the great King of the Mongols" is an allusion to Genghis Khan, we should consider that by the year 1227 the Mongol empire of Khan stretched from the Pacific to the Black Sea. However, if the phrases "the great King of the Mongols" and "the great King of Terror" were both intended as allusions to present or future Asian military leaders, then there are many possibilities to consider.

It is also a possibility that "great King of the Mongols" is an allusion to the Mongol race (rather than a specific reference to the nation of Mongolia). The Chinese, Mongolians, Siberians, (and other oriental Asians) are all categorized as being members of this ethnic division of the human race. So in consideration of this fact, we could easily hypothesize that "the great King of the Mongols" is an allusion to China (the most economically and militarily powerful Mongol nation in 1999). But could a war between North Korea and South Korea be a possibility? They have been sworn enemies for four decades. Could there be more trouble in Iraq? Could there be trouble somewhere else in the Middle East or Europe? Could a future war between communist China and democratic Taiwan be referenced here? Could a war between India and Pakistan be a possibility here? Now that each of these two nations possess (sic) *nuclear weapons the tension in Asia is rising. Since July (or September) of 1999 is near the change of the millennium, and quatrains #2-46 and #10-74 suggest that warfare might possibly occur near the time of this change, one could wonder if these three quatrains could be connected with each other?*

As of March of 1999, NATO nations began a grand aerial bombing campaign against the Yugoslavian military in Kosovo and Serbia. Although this event began four months before July (the "seventh" month), many of the parameters surrounding it appear to fulfill parts of this unique prophecy. In example, since this quatrain mentions war "before" July of 1999, would NATO's military campaign in Yugoslavia (which ended in June) qualify as the event referencing warfare before July of 1999? In May, NATO air forces accidentally bombed the Chinese Embassy in Belgrade. Could this event be a reference to "bringing back to life the great king of the Mongols" (perhaps alluding to upsetting the Chinese)?

THE 2000 POST FACTO VIEW: Who, or what, was the great King of Terror (who came from the sky in July of 1999)? Was it an allusion to a person, or to something else? Over the last few decades this phrase has generally been interpreted as a reference, or allusion, to a human being (usually to a terrorist or Asian war lord). But that was speculation, and or assumption, on behalf of most Nostradamus authors (including myself).

Could the crash of John F. Kennedy Jr's airplane in July of 1999 fulfill the line "in the seventh month from the sky will come the great King of Terror"? Could the human fear of death and bodily injury be the intended definition of "the great King of Terror"? It might be possible! Since Nostradamus made several predictions about the death of President Kennedy (and Robert F. Kennedy), could a prediction about the death of John F. Kennedy Jr. fit into a familiar Nostradamian theme? It would certainly seem possible.

of the moon is abundantly more craterous than the side that faces Earth". Still, the hypothesis was in-principle testable in that one could imagine a day when a space ship would travel to or around the moon and confirm or disconfirm this hypothesis. Last, to say a hypothesis is articulated either "with regard" or "without regard" for the four other criteria is to say something fairly vague. What I mean here is that to be at least marginally scientific a hypothesis should aim to fit as many of these other criteria as possible in the best possible manner.

Let us now turn to three illustrations.

Freud on the Historical Roots of the Oedipus Complex

Freud was never shy in letting it be known that the Oedipus complex—specifically, the unconscious desire for each boy to kill his father and take for himself his mother—was the cornerstone of his psychoanalytic theory. So crucial it was to psychoanalysis that he sought a historical explanation for it, rooted in precivilized, human-horde behavior. *Totem and Taboo* (1913) was his first attempt to disclose this origin. In tracing out his thoughts on the roots of the Oedipus complex, I draw chiefly from *Totem and Taboo, Civilization and Its Discontents,* and *Moses and Monotheism.*

Freud's Primal-Horde Hypothesis

Human guilt (i.e., the super-ego) goes back to a killing of a primal father in the early history of humans. Before going into the details of the story, I list some of Freud's assumptions.

- ***NOTHING ONCE COMMITTED TO MEMORY CAN EVER PERISH.*** *"In mental life nothing which has once been formed can perish—that everything is somehow preserved and that in suitable circumstances ... it can once more be brought to light".*[1]
- ***THE PSYCHOLOGICAL DEVELOPMENT OF AN INDIVIDUAL IN A LIFETIME IS THE SAME AS THE PSYCHOLOGICAL DEVELOPMENT OF THE SPECIES OVER TIME (ONTOGENY RECAPITULATES PHYLOGENY).*** *"No one can have failed to observe, in the first place, that I have taken as the basis of my whole position the existence of a collective mind, in which mental processes occur just as they do in the mind of an individual. In particular, I have supposed that the sense of guilt for an action has persisted for many thousands of years and has remained operative in generations which can have had no knowledge of that action".*[2] *"Although the majority of human children individually pass through the Oedipus complex, yet after all it is a phenomenon determined and laid down for him by heredity, and must decline according to schedule when the next pre-ordained stage of development arrives".*[3]
- ***HAPPINESS IS NOTHING OTHER THAN PSYCHO-SEXUAL RELEASE.*** *"What we call happiness in the strictest sense comes from the (preferably sudden) satisfaction of needs which have been damned up to a high degree, and it is from its nature only possible as an episodic phenomenon".*[4]

In Totem and Taboo, *Freud begins by defining both "totem" and "taboo". A totem is an animal, plant, or even natural phenomenon (like rain or water) that takes on the identity of a common ancestor of a clan—a guardian spirit or helper. Clansmen are under a sacred obligation not to kill or destroy their totem and to avoid eating its flesh (or deriving benefit from it in other ways). He writes, "In almost every place where we find totems we also find a law against persons of the same totem having sexual relations with one another and consequently against their marrying".*[5] *A taboo he defines as follows:*

> *Taboo is a primaeval prohibition forcibly imposed (by some authority) from outside, and directed against the most powerful longings to which human beings are the subject. The desire to violate it exists in the unconscious; those who obey the taboo have an ambivalent attitude to what the taboo prohibits.... The fact that the violation of a taboo can be atoned for by a renunciation shows that renunciation lies at the basis of obedience to taboo.*

The two laws (i.e., taboos) of totemism are then:

> ***DO NOT KILL THE TOTEM ANIMAL.*** *(This is founded emotionally; the killing of the primal father was committed and could not be undone.)*
>
> ***THERE ARE TO BE NO SEXUAL RELATIONS WITH WOMEN IN SAME TOTEM.*** *(Sexual relations divide a community).*

These have risen to the status of taboos, because everyone has powerful impulses to do them.

Freud then works back to what he calls the "primal horde", which existed at a point in time that cannot be geologically established.[6] *In such a horde, the leader, its "father", was the totem and all-powerful. Following Darwin's observations of animal behavior, where the dominant male takes all the females for himself to the frustration of the lesser males, Freud argues that the sons, through realization of strength in numbers, dealt with their sexual frustration by killing the primal father and eating his flesh. The sons banding and acting together was the first true social action of the first social entity. Freud adds, "The cannibalistic act thus becomes comprehensible as an attempt to assure one's identification with the father by incorporating a part of him".*[7] *Yet, since the primal father was also seen as the totem, the sons violated the first totemic law in killing the father. Thus totemic religion, the father of all religions, was founded.*

> *Totemic religion arose from the filial sense of guilt, in an attempt to allay that feeling and to appease the father by deferred obedience to him. All later religions are seen to be attempts at solving the same problem.... Totemic religion not only comprised expressions of remorse and attempts at atonement, it also served as a remembrance of the triumph over the father.... Society was now based on the sense of guilt and the remorse attaching to it; while morality was based partly on the exigencies of this society and partly on the penance demanded by the sense of guilt.*[8]

Freud arrived at this conclusion following the principle of recapitulationism—the notion that the development of an individual in time retraces the development of the species of humans in time, and conversely. In short, Freud believed that the species itself must have gone through its own Oedipal tension and resolution. The guilt upon killing the primal father was so overwhelming that the deed was completely forgotten and the primal father was made a god. This deed was the origin of god the father, repression, and the super-ego.

> *The psychoanalysis of individual human beings, however, teaches us with quite special insistence that the god of each of them is formed in the likeness of his father, that his per-*

sonal relation to God depends on his relation to his father in the flesh and oscillates and changes along with that relation, and that at bottom God is nothing other than an exalted father figure.[9]

Freud finds additional confirmation that such a deed must have occurred in the notions of "god the father" and "origin sin" in the Christian faith. He writes:

There can be no doubt that in the Christian myth the original sin was one against God the Father. If, however, Christ redeemed mankind from the burden of original sin by the sacrifice of his own life, we are driven to conclude that the sin was a murder. The law of talion *(i.e., an eye for an eye), which is so deeply rooted in human feelings, lays it down that a murder can only be expiated by the sacrifice of another life: self-sacrifice points back to blood-guilt. And if this sacrifice of a life brought about atonement with God the Father, the crime to be expiated can only have been the murder of the father.*[10]

He sums, "the Christian communion has absorbed within itself a sacrament which is doubtless far older than Christianity".

Of the indispensability of the Oedipus complex, he states:

At the conclusion, then, of this exceedingly condensed inquiry, I should like to insist that ... the beginnings of religion, morals, society and art converge in the Oedipus complex. This is in complete agreement with the psycho-analytic finding that the same complex constitutes the nucleus of all neuroses, so far as our present knowledge goes. It seems to me a most surprising discovery that the problems of social psychology, too, should prove soluble on the basis of one single concrete point—man's relation to his father.[11]

Assessment of **Freud's Primal Horde Hypothesis**

Model/ Theoretical Hypothesis

M: The historical model of the Oedipus complex.
H_O: The Oedipus complex has it biological roots in the killing of the primal father.

Prediction

It is hard to see what prediction (here, retrodiction) one can deduce from such a claim.

Background Information/Auxiliary Hypotheses

* Freud's notion of happiness as sudden release of sexual impulses.
* Charles Darwin states that men in primitive times lived in small hordes, under the domination of one strong male, who is responsible for the sexual frustration of all other males in the horde.

* Recapitulationism—the view that the development of an individual retraces the development of the species—is assumed true.
* The notion that no deed, whether that of an individual or that of an early ancestor, is ever completely forgotten. The latter involves the notion of a collective unconscious.
* Freud's notion that powerful prohibitions must be directed against powerful impulses to act in certain ways (i.e., to kill or commit incest).

Results

Freud's reference to Darwin's observations of primal human hordes.

Analysis

Freud's model seems completely speculative. There are no data to speak of, other than his reference to Darwin's data, which must be taken analogically. Everything else seems to be deduced from certain assumptions listed in the background information. There seems to be no way to test for this hypothesis. This alone is sufficient to show it is pseudo-scientific.

In addition, and the rest here is superfluous, the hypothesis does not meet the criteria of simplicity and fruitfulness. First, there is nothing simple about this hypothesis; it seems entirely unneeded. Moreover, as it does not straightforwardly suggest new lines of research, it is not fruitful. Furthermore, in many places the reasoning borders on absurdity. For instance, Freud combines Christ's dying on the cross with the law of *talion* (i.e., an eye for an eye) to conclude that the sin for which Christ was put to death must have been a murder. This he takes to be additional confirmation of his thesis. Certain of the auxiliary hypotheses are quite questionable as well (e.g., his definition of happiness, the recapitulation thesis, the existence of a collective unconscious), all of which are needed for his account to be true. The model does seem to meet the criteria of scope and conservatism. It has great scope (explanatory power), but at expense of touching base with reality. It meets the demand for conservatism *only* insofar as it is driven by the phylogenetic principle, assumed true at the time but now known to be false, and by observations of animal behavior, which are not wholly correct.

Conclusion

With no data having a bearing on the hypothesis and no straightforward prediction being put to the test, it is impossible to regard this as a scientific hypothesis. Here Freud is engaged in pseudo-science.

Chiropratic

From 1970 to present, the number of practicing chiropractors in the United States increased from 13,000 to over 50,000. This increase shows that chiropractic is becoming a larger part of the American health-care system. Does chiropractic work? The answer to this question is, in large part, in the claims made on behalf of chiropractic.

Alternative-Medicine Practitioner Refuses Alternative Method of Payment (*The Onion*, March 29, 2006, 42.13)

Alternative-medicine practitioner Annabeth Severin, a Portland-area acupuncturist and holistic healer, announced Tuesday that she is refusing to accept anything but conventional monetary compensation from her patients. "I'm sorry, but there just isn't any sound economic theory to support the idea that bartering or visualization of payment has the same effect as traditional cash or check up front", Severin said. Her customers are protesting her billing methods, saying that removing money from their accounts would be financially invasive and spiritually upsetting to their karmic and bank balances.

Efficacy of Chiropractic Care

The following is an article on the scope and aims of chiropractic care from the web page of the American Chiropractic Association.[12]

Chiropractic Philosophy

As a profession, the primary belief is in natural and conservative methods of health care. Doctors of chiropractic have a deep respect for the human body's ability to heal itself without the use of surgery or medication. These doctors devote careful attention to the biomechanics, structure and function of the spine, its effects on the musculoskeletal and neurological systems, and the role played by the proper function of these systems in the preservation and restoration of health. A Doctor of Chiropractic is one who is involved in the treatment and prevention of disease, as well as the promotion of public health, and a wellness approach to patient healthcare.

Scope of Practice

Doctors of Chiropractic frequently treat individuals with neuromusculoskeletal complaints, such as headaches, joint pain, neck pain, low back pain and sciatica. Chiropractors also treat patients with osteoarthritis, spinal disk conditions, carpal tunnel syndrome, tendonitis, sprains, and strains. However, the scope of conditions that Doctors of Chiropractic manage or provide care for is not limited to neuromusculoskeletal disorders. Chiropractors have the training to treat a variety of non-neuromusculoskeletal conditions such as: allergies, asthma, digestive disorders, otitis media (non-suppurative) and other disorders as new research is developed.

A variety of techniques, treatment and procedure are used to restore healing which will be the topic of future education releases.

Assessment of **Efficacy of Chiropractic Care**

Model/Theoretical Hypothesis

M: The chiropractic model of physical health.

H_{CM1}: Chiropractic care treats *and* prevents neuromusculoskeletal disorders (headaches, joint pain, neck pain, low-back pain, and sciatica) and disease (allergies, asthma, digestive disorders, non-suppurative *otitis media* and other disorders as new research is developed).

H_{CM2}: Chiropractic promotes both public health and a wellness approach to patient healthcare.

Prediction

P_{CM1}: If the model is true, then those who visit a chiropractor for neuromusculoskeletal disorders and certain diseases should experience not only relief from symptoms, but also overall improved health through spinal manipulation.
P_{CM2}: If the model is true, then those who visit a chiropractor for preventative maintenance should experience overall improved health through spinal manipulation.

Results

(Notice below how much "outside work" I have done on chiropractic before assessing the claims on behalf of it that are listed on the ACA webpage.)

* The *New England Journal of Medicine* says spinal manipulation may be no more effective than physical therapy in the treatment of back pain and only marginally more effective than a self-help instruction booklet.[13]
* There are no scientific studies to indicate that vertebral misalignment or any other problem in the spine is a cause of disease or infection. For instance, chiropractic used to treat asthma in children has not been shown to be effective.[14]
* A 1996 RAND report estimated that stroke and other injuries can result from cervical spine manipulation through arterial dissection.[15]
* Regular manipulation can cause problems requiring continued manipulation and result in more harm than good.[16]

Analysis

The model claims, via its fairly clear predictions, that chiropractic can cure and prevent disease. Those claims are not substantiated by research. That chiropractic contributes to overall bodily health is questionable as well, as chiropractic has, in some cases, resulted in impaired physical health. Here also the evidence is inconclusive. Consequently, the predictions made by chiropractic, though unambiguous, seem to be false, and that is fairly damning.

The model seems simple enough. That perhaps is its main virtue. It is also seems fruitful. Chiropractic makes an abundance of claims for overall health—each of which is testable, though many, if not most, of which are probably false. It fails conservatism miserably however, as much evidence in conventional medical practice suggests overall health and disease control go far beyond neuromusculoskeletal care. Last, it does have sufficiently large scope over human health, though the claims it makes are probably highly exaggerated.

Conclusion

Chiropractic seems a legitimate means of treating some cases of lower-back pain, but its use as a means of disease control and overall preventative health is highly suspect. Given the wide claims that are made on its behalf and the lack of evidence to support them, chiropractic may be categorized as science, but it is shoddy science.

Homeopathy

Frustrated with conventional medical approaches and their tendency to treat the problem and not the whole person, alternative approaches to medicine are enjoying great popularity. Homeopathic treatment is one such alternative and it is growing at a yearly rate of over twenty-five percent.

What is homeopathy and does it work? To answer these questions, let us look at the origin of this approach and some of its underlying assumptions.

An Effective Alternative Approach to Medicine?

German physician Samuel Hahnemann (1755-1843) is said to be the founder of homeopathy. As a doctor, he became dissatisfied with the state of medicine at his time. He stopped practicing and began to translate medical and chemical texts, which led to his views on homeopathic remedy.

Hahnemann believed that diseases had non-material causes. He wrote:

> *The causes of our maladies cannot be material, since the least foreign material substance, however mild it may appear to us, if introduced into our blood-vessels, is promptly ejected by the vital force, as though it were a poison...no disease, in a word, is caused by any material substance, but that every one is only and always a peculiar, virtual, dynamic derangement of the health.*[17]

Treatment, he believed, should consist of remedies that are similar to the disease itself. As polio is treated with a polio vaccine, other diseases should be treated similarly. For instance, if symptoms are linked with mercury poisoning, then mercury would be the cure.

Hahnemann supported his intuitions by "proving" them on himself and other adherents of his views. He and his assistants ingested small amounts of nearly 100 substances. Tthey then noted note any reactions to any particular substance. As some of these substances (e.g., snake venoms) were dangerous, doses were extraordinarily small.

Homeopathy works according to the principle of superdilution, which states that the potency of a remedy is in proportion to its dilution. Remedies are diluted by a factor of 10 (D) or 100 (C). A D1 dilution of particular snake venom is one part venom and 10 parts water. A D2 dilution is one part of D1 mixed with 10 parts of water, and so forth. It is easy to see how rapidly the venom is diluted by continuing this process. Consider a C20 dilution of the venom (one part venom and 100 parts water diluted 20 times). Such a superdilution would likely not contain even one molecule of the poison, and such superdilutions are claimed to be maximally potent remedies.

Surprisingly, this type of reply was not too troubling to Hahnemann. As the causes of disease were not material, so too the cures were not material. Superdilution was thought (somehow) to bring out certain vital, non-material powers in the solution. These vital powers were then stirred up through "potentization". Writes Hahnemann, "Homeopathic potentizations are processes by which the medicinal properties of drugs, which are in a latent state in the crude substance, are excited and enabled to act spiritually upon the vital forces".[18] *At the end of the every process of dilution, the solution was "succussed" (shaken vigorously, for liquids) or "triturated" (ground up vigorously, for powders) so as to liberate the vital powers of the substance.*

Hahnemann believed, following ancient Greek medicine,[19] *that there were four main body types following the four humors—bile (choleric: easily angered and moved); black bile (melancholic: sad), blood (sanguine: volatile and warm-hearted) and phlegm (phlegmatic: sluggish and apathetic)—and that homeopathic remedy must follow these bodily constitutions.*

The wide use of homeopathic remedy is almost always used as an argument for its effectiveness. Yet, homeopathists claim more. It is extraordinarily effective, completely safe, non-addictive, and natural way to cure disease.

> *When the correct remedy is taken, results can be rapid, complete and permanent.... It treats all the symptoms as one, which in practical terms means that it addresses the cause, not the symptoms. This often means that symptoms tackled with Homeopathy do not recur.*[20]

There is little unambiguous evidence for its effectiveness. David Ramey et al write:

> *Subsequent to submission of the most recent meta-analysis in human medicine, several good trials of homeopathic medications have been conducted in human medicine. Randomized, placebo-controlled double blind studies have shown homeopathic remedies to be ineffective in the treatment of adenoid vegetations in children, for control of pain and infection after total abdominal hysterectomy and for prophylaxis of migraine headache. Furthermore, to date, no single study of homeopathy, showing positive results, has been successfully replicated.*[21]

There have, of course, been many reports of homeopathic cures of diseases. I list one such study below, taken directly from *The Journal of Alternative and Complementary Medicine*.

When Conventional Treatment Is Not Enough: A Case of Migraine without Aura Responding to Homeopathy[22]

Following three years of unsuccessful conventional treatment, a 55 year old male suffering from common migraine which would commence with nausea followed by vomiting every hour for 12 hours and throbbing pain well localized to the left fronto-parietal area, was referred to Glasgow Homeopathic Hospital. Consultation with a homeopathic physician, who also has extensive experience in diagnosis and treatment of headache disorders, leads to the prescription of a single homeopathic remedy (Bronia) which was absolutely effective for the condition. On follow-up 2 months later, the patient has been headache-free and had lost no time from work. He had only taken the Bronia for 3 weeks (i.e.; 12 doses). He remains attack-free three years after treatment. This case is offered as an open, admittedly retrospective study, comparing the best of conventional migraine therapy with appropriate homeopathic therapy in the same patient.

***Assessment of* An Effective Alternative Approach to Medicine?**

I focus here on the "study" reported in *The Journal of Alternative and Complementary Medicine.*

Model/Theoretical Hypothesis

M: The homeopathic model of curing disease.
H_{HM}: A superdiluted bronia solution cures migraine headaches.

Prediction

If H_{HM} is true, then Bronia will cure the subject of migraine headaches.

Results

A 55-year-old male suffers from common migraine headaches, accompanied by nausea. The subject takes 12 doses of Bronia over a period of three weeks. The man has been headache-free for three years.

Background Information

Ramey et al's claim, "[T]o date, no single study of homeopathy, showing positive results, has been successfully replicated".

Analysis

This may be a case of *single-shot inductivism*, which judges a remedy effective from one report of cure in a single case. If so, the argument is dismally weak. If not, lack of experimental controls makes it impossible to know that the homeopathic remedy is the cause of the lack of migraines in three years. No other causal factors are considered and ruled out. A simpler explanation is that the placebo effect is at work. No "study" such as this could get published in a legitimate medical journal.

Overall, Ramey et al's claim about the irreplicability of homeopathic studies suggests most are conducted sloppily, with few controls. This shows the data on homeopathy are inconsistent.

The assumption of superdilution in homeopathy is absurd and goes against everything we know about potency. Thus, it fails conservatism. Again, as the powers of the superdiluted substance are *immaterial*, there can be no tests done to confirm that they have such powers.

Conclusion

Homeopathy and homeopathic treatment are pseudo-scientific.

Key Terms

criteria of adequacy
ad hoc hypothesis
fruitfulness
conservatism
marginally scientific
roots of Oedipus complex
homeopathy
testability
simplicity
scope
scientific
pseudo-scientific
recapitulationism
single-shot inductivism

Text Questions

* Why are the criteria of adequacy, with the exception of testability, not meant to be taken as necessary criteria for proper scientific practice?
* Can you think up a scenario where any two of the criteria of adequacy might come into conflict?
* What mistakes did Freud make in trying to give his Oedipus complex historical validity? Does that mean we must reject the Oedipus complex as a psychodynamic (ontogenetic) phenomenon?
* Does labeling chiropractic a "marginally scientific" discipline mean that it is senseless to see a chiropractor for any bodily disorders?
* Why do you think Hahnemann said that the superdiluted powers of his remedies had immaterial powers? Why does dilution itself, let alone superdilution, violate the criterion of conservatism?

Text-Box Questions

* Why does John Losee call philosophy of science a "second-order criteriology"?
* What sorts of "red flags" are there that make you suspicious of the "interpretation" of Quatrain X.72? Can you come up with your own interpretation—equally "accurate" and eerie?

---

1 Sigmund Freud, *Civilization and Its Discontents,* trans. James Strachey (New York: W.W. Norton & Co., Inc., 1961).
2 Sigmund Freud, *Totem and Taboo*, trans. James Strachey (New York: W.W. Norton & Company, 1962).
3 Sigmund Freud, "The Passing of the Oedipus Complex", 1924.
4 Sigmund Freud, *Civilization and Its Discontents.*
5 Sigmund Freud, *Totem and Taboo.*
6 Sigmund Freud, *Moses and Monotheism*, trans. Katherine Jones (New York: Vintage Books, 1967).
7 Sigmund Freud, *Moses and Monotheism.*
8 Sigmund Freud, *Totem and Taboo.*
9 Sigmund Freud, *Totem and Taboo.*
10 Sigmund Freud, *Totem and Taboo.*

---

11 Sigmund Freud, *Totem and Taboo*.
12 http://www.amerchiro.org/media/whatis/history_chiro.shtml.
13 D. C. Cherkin et al., "A Comparison of Physical Therapy, Chiropractic Manipulation, and Provision of an Educational Booklet for the Treatment of Patients with Low Back Pain", *New England Journal of Medicine* 339:1021-1029, 1998.
14 J. P. Balon et al., "A Comparison of Active and Simulated Chiropractic Manipulation as Adjunctive Treatment for Childhood Asthma", *New England Journal of Medicine* 339:1013-1020, 1998.
15 I. E. Coulter et al., "The Appropriateness of Manipulation and Mobilization of the Cervical Spine" RAND Report (Santa Monica, CA: RAND, 1996) 18-43.
16 Samuel Homola, D.C., "Chiropractic: Does the Bad Outweigh the Good?", *Skeptical Inquirer*, Jan/Feb, 2001.
17 Samuel Hahnemann, *Organon of Medicine*, 6th ed. (Calcutta: M. Bhattacharyya & Co.; 1960).
18 Samuel Hahnemann, *The Chronic Diseases* (New York, 1846).
19 Beginning around the fifth century B. C.
20 http://www.abchomeopathy.com/homeopathy.htm, 1.
21 David Ramey et al. "Homeopathy and Science: A Closer Look", http://spot.colorado.edu/~vstenger/Medicine/Homeop.html.
22 *The Journal of Alternative and Complementary Medicine*, 1997 1:2

Module 26
Exercises for Science, Marginal Science & Pseudo-Science

"I am ... accustomed to reaffirm with emphasis my conviction that the sun is real, and also that it is hot—in fact hot as Hell, and that if the metaphysicians doubt it they should go there and see". Winston Churchill, *My Early Life*

EVALUATE EACH OF THE EXAMPLES BELOW according to the six guidelines for models below.

MODEL/THEORETICAL HYPOTHESIS: *What is the model under investigation? What theoretical hypotheses drive this model?*
PREDICTION: *What prediction(s) does the model through any of its theoretical hypotheses make?*
BACKGROUND INFORMATION/RESEARCH METHODS: *What auxiliary hypotheses here come into play? Is there relevant background information? Is there anything worth mentioning about the research methods?*
RESULTS: *What are the relevant data?*
ANALYSIS: *Are the data consistent with the prediction?*
If so, to what extent do they confirm it? Are there other models or hypotheses consistent with the data that cannot be ruled out?
If not, must the model be rejected categorically or is there some non-ad-hoc way to salvage it?
CONCLUSION: *Give your verdict of the model here succinctly and precisely.*

Use outside resources, like web resources, to glean further information for your research, if necessary. In addition, use the five criteria of adequacy (see Module 25) in your analysis.

Carl Jung and the Interpretation of Dreams

Carl Jung and Number Dreams

In an essay entitled "On the Significance of Number Dreams", psychologist C.J. Jung offers his insights on how free association can be used to shed light on numbers in dreams. For Jung, following Freud, free association is an analytic method that allows patients, with the help of a qualified analyst, to work back from key concepts, here in a dream, to underlying tensions causally responsible for them. In this essay, Jung gives several examples, of which I list one below. It is the dream fragment of a mid-aged man, who was troubled at the time of the dream by an extramarital love-affair.

> *The dreamer shows his season ticket to the conductor. The conductor protests at the high number on the ticket. It was 2477.*

Free Association

Free association is a psychoanalytic method, made famous by Freud, for disclosing unconscious motivations underlying observable behaviors. For Freud, the Unconscious is a warehouse, as it were, of memories so painful that they have been barred from Consciousness. These unconscious memories, as repressions, still seep into daily life through slips of the tongue, dreams, and jokes—all of which may be taken as everyday-life manifestations of hysteria.

Freud himself writes about the impact of Bernheim's work on hypnosis that led to the development of his method of free association. He says, in a 1922 essay entitled "Psychoanalysis":

> The abandonment of hypnosis seemed to make the situation hopeless, until the writer recalled a remark of Bernheim to the effect that things that had been experienced in a state of somnambulism were only apparently forgotten and that they could be brought into recollection at any time if the physician insisted forcibly enough that the patient knew them. The writer therefore endeavored to press his unhypnotized patients into giving him their associations, so that from the material thus provided he might find the path leading to what had been forgotten or warded off. He noticed later that such pressure was unnecessary and that copious ideas almost always arose in the patient's mind, but that they were held back from being communicated and even from becoming conscious by certain objections put by the patient in his own way. It was to be expected—though this was still unproved and not until later confirmed by wide experience—that everything that occurred to a patient seeking out from a particular starting-point; hence arose the technique of educating the patient to give up the whole of his critical attitude and of making use of the material which was thus brought to light for the purpose of uncovering the connections that were being sought. A strong belief in the determination of mental events certainly played a part in the choice of this technique as a substitute for hypnosis.

In short, early failures to get information from patients were overcome by pressuring his patients to say anything that came to mind, however trivial it might seem. Freud soon realized that patients were resisting therapy and he could overcome this resistance slowly, over time, through analytic sessions where he merely coaxed them to associate freely.

In this manner, Freud thought free association was a method for retracing, from etiological symptoms, the tensions (usually sexual in nature) causally responsible for them. He did not believe free association was an infallible method, but merely the best method available to the analyst.

Against Freud's method of free association, Adolf Grunbaum in *The Foundations of Psychoanalysis* has argued that the sole justification that free association is a viable method for disclosing underlying symptoms is the therapeutic success of the analyst in clinical practice. It is just this that clinical "successes" cannot guarantee. He sums:

> Without this vindication or some other as yet unknown epistemic underpinning, not even the tortures of the thumbscrew or of the rack should persuade a rational being that free associations can certify pathogens or other causes! For, without the stated therapeutic foundation, this epistemic tribute to free associations so far rests on nothing but a glaring causal fallacy.

Grunbaum concludes that the method of free association, as a method for securing the causes underlying hysterical symptoms, is therefore methodologically unreliable. The power of suggestion by the analyst can never be ruled out.

Freud himself however was not unaware of the difficulties involved in using free association as a valid method to "lift" underlying repressions. In a late essay, "Constructions in Analysis", Freud says rather smugly:

> The danger of our leading a patient astray by suggestion, by persuading him to accept things which we ourselves believe but which he ought not to, has certainly been exaggerated. An analyst would have to behave very incorrectly before such a misfortune could overtake him; above all, he would have to blame himself with not allowing his patients to have their say. I can assert without boasting that such an abuse of "suggestion" has never occurred in my practice.

Jung notes that the analysis of the dream showed that the normally generous dreamer was troubled by the expenses of his love-affair and unconsciously was trying to end the affair. Thus, Jung begins on the assumption that the number "2477" had "a financial significance and origin". An estimate of the expenses led to 2387 francs—a number close to but not equivalent to 2477. So Jung turned to free association. He writes:

> *It occurred to him that in the dream the number appeared divided: 24 77. Perhaps it was a telephone number. This conjecture proved incorrect. The next association was that it was the sum of various other numbers. At this point the patient remembered telling me earlier that he had just celebrated the hundreth birthday of his mother and himself, since she was sixty-five and he was thirty-five. (Their birthdays fell on the same day.)*[1]

Thus, free association led to the following analysis:

He was born on (day.month):	*26.II*
His mistress:	*28.VIII*
His wife:	*1.III*
His mother:	*26.II*
His 1st child:	*29.IV*
His 2nd child:	*13.VII*
He was born on (month.year):	*II.75*
His mistress:	*VIII.85*
He was now:	*36*
His mistress:	*25*

Jung then proceeds to add these together:

$$262 + 288 + 13 + 262 + 294 + 137 + 275 + 885 + 36 + 25 = 2477$$

This series, Jung states, shows that there is a deeper significance to the dream. The series of numbers, including all the members of his family, shows his attachment to his family. This attachment was causing problems, at least unconsciously, with his relationship to his mistress.

Jung states that examples such as this, even though they may float "on a sea of uncertainties", offer at least a preliminary analysis for luckier investigators at some future time.

Cold Reading

Psychics on Reading People's Minds

A female client enters the psychic's studio and takes a seat. The psychic notices that the woman is excessively obese and wears an inordinate amount of makeup that is hastily applied. She is garishly dressed and sports a cheap bracelet with a bold letter "J" on her left wrist as well as an engagement ring. She fidgets in her seat nervously and has a difficult time looking the psychic in the face. After exchanging some pleasantries, the psychic begins the session as follows:

P: I feel that something important is weighing you down—keeping you up at night, causing you to take frequent naps during the day.
C: I sleep a lot. I sleep all the time. At night and during the day. After work especially.
P: I see a very important man in your life. (A long pause here where the C says nothing.) Perhaps your husband?
C: I'm not married. (C looks down.)
P: Yet I see marriage. I see you as married. I hear a name too. "Jim" or "John" or...
C: How did you know? Jerry is my fiancé. At least, he was.
P: Yes, yes! I said "Jim" or "John". I wasn't sure. I was really hearing a man's name—certainly the name of a man—a man with the letter "J". Hm-mmm. I see now that you even have a bracelet with a "J" on it.
C: He left me two weeks ago. (C's eyes well up with tears.) We were supposed to get married.
P: I see that he has hurt you deeply. (P pauses.) I see a truck. Is he a truck driver?
C: He doesn't have a job now. He was helping out a friend for a while. (C pauses.) He used to drive a truck, I think. Years ago ... before he met me. I'm not really sure of this though, but I think you're right.
P: I could see this. Yes, I see a truck in his life. Hmmm, there's more. There's something important about a truck in his life even now. It's not important to go into this now.... But I do see clearly, very clearly, that he's far away from you.
C: Well ... he hasn't moved away—at least, not far away. He's back with his ex. He's living with her. (C begins to cry.)
P: Yes, of course! He is *far away from you. That's what I see! He's far away emotionally. He's far from your heart. He's torn your heart! He's far away from you emotionally, and you still love him ... a lot. (C cries more.) You're deeply hurt—this came to you unexpectedly, I see. Yet I see great resolve, great strength in you. You've been hurt before.*
C: Many, many times! Too many times! I'm so tired of it all! He's an ass! (More crying.)
P: (A very long pause here.) You've come here for answers, haven't you?
C: I just want to know if he'll come back? Don't even know ... don't know if, if I want him back....
P: He'll come back. I see this clearly. But I see a different path for you. An opportunity. Another man. He's very handsome. He'll make you very happy. I see excitement in your life, like never before—happiness like never before. But, before this, I sense that you must handle the pain bravely. If you don't, you may never see the opportunity. You can handle the pain bravely. You're a woman of great strength!

Psychic Helps Police Waste Valuable Time, *The Onion*, March 24, 2004, 40.12

More than 36 hours after the disappearance of 13-year-old Heather Jordan, Manchester police hired local psychic Lynette Mure-Davis to help waste their valuable time Monday. "I see a river... and along the banks is an outcropping with five lilac bushes", said Mure-Davis, who then paused a full 90 seconds to "collect vibrations" from Jordan's scarf. "I also see a man...tall, but stocky, wearing...a hat. And an animal, perhaps a dog". As of press time, Jordan was still trapped under a collapsed utility shed three blocks west of her house.

Organisms, Orgasms, and Orgone Energy[2]

Medical Orgone Therapy

Medical orgone therapy is a unique approach to treating emotional and physical illness by reducing or eliminating those barriers that block the natural expressions of emotion and healthy sexual feeling. This method of treatment was developed by Austrian psychiatrist and scientist Wilhelm Reich (1897-1957). After years of clinical and experimental laboratory research, Reich concluded that emotions, sexual feelings, and all life processes are expressions of a biological energy in the body. He further concluded that this life energy is related to bioelectricity but is fundamentally different. He called this energy "orgone energy". Reich theorized that orgone energy fills the universe and pulsates in all living things. He believed that deep, genuine love and the ability to experience a gratifying orgasm mutually with one's partner are the fullest and deepest expressions of our being and are central to maintaining optimal health.

Reich contended that in almost all individuals, the flow and release of orgone energy are blocked by chronic muscle contractions in various areas of the body and by emotional attitudes adopted early in life. The "nice little girl" who never gets angry, and the "strong, brave boy" who never shows fear and sadness are but two examples of such attitudes that prevent the full and rational expressions of natural emotions. Blocked emotions interfere with pleasure in the life and cause sexual feelings to become disconnected from tender emotions of love. Without emotional release, anxiety develops, which further increases physical and emotional contraction. This cycle results in a range of problems such as feelings of emptiness, depression, irrational fears, and self-destructive behavior. Medical orgone therapy employs direct work on the body, especially on spastic muscles, with verbal therapy to bring about a healthy state accompanied by satisfaction in one's work and love life.

The History of Medical Orgone Therapy

Dr. Wihelm Reich began his career as a student and colleague of Sigmund Freud, the founder of psychoanalysis. During his work as a psychoanalyst, Reich discovered that the individual's deep emotions were bound up in defensive character attitudes, which he called "character armor". To treat these problems Reich developed a highly effective technique of character analysis, still used today by other psychotherapies, which focuses on the individual's attitudes and present-day concerns and less so on past relationships within the family.

While working with patients, Reich observed that only those who developed a satisfactory, healthy sexual life fully resolved their neurotic symptoms. Satisfaction was not determined by the mere presence of sexual activity. Rather, it required the ability to give in to both deep, tender love feelings and the intense sexual sensations that are experienced with a total body orgasm. Reich asserted that this all encompassing experience is possible only with a partner of the opposite sex, and concluded that only such a complete orgasm could regulate the energy metabolism.

Reich observed that defensive emotional attitudes are not just "in the mind" but are held in the body's muscles, and he called this the "muscular armor". He also found that inhibited feelings were accompanied by restrictions in respiration. These realizations let him to the ground-

breaking conclusion that the successful treatment of emotional problems requires work on the body combined with verbal therapy....

The Theory of Medical Orgone Therapy

Medical orgone therapy is based on Wilhelm Reich's theory of armoring. The method of treatment developed from observations on the movement and blockage of energy in the body. Infants and children naturally feel pleasure and reach out to the world. If these impulses are frustrated, the child contracts and develops methods to adapt to the stress. If frustrations continue, these defensive reactions become chronic and extend into adult life even when they are no longer needed. For example, a child may develop a submissive manner to deal with an angry parent and then as an adult react submissively to all authority figures, even when it would be better to be assertive. Other common examples of "character armor" are found in individuals who present themselves as aloof, superior, "cool", sophisticated, cute, or special. Medical orgonomists consider these character attitudes to be manifested in actual muscular rigidities (muscular armor), which hold back intolerable or unacceptable emotions such as anger, fear, or sadness. Individuals are usually unaware of their muscular armor or that their physical problems, such as headaches, stiff neck, or back pain, are often rooted in repressed emotions.

Armoring forms in infancy and early childhood as defense against painful feelings, but it is not a satisfactory solution because it later interferes with healthy emotional life and energy discharge. Medical orgone therapy strives to eliminate chronic armor to restore the individual to more natural functioning in all aspects of his or her life.

Medical Orgone Therapy in Practice

The medical orgonomist is trained to understand the patient in all respects and diagnose the patterns of character and physical armor. Because all medical orgone therapists are physicians who also have specialized training in psychiatry, they are equipped to diagnose physical conditions and work directly on their patients. Treatment includes character analysis and the release of buried emotions facilitated by breathing and direct work on spastic muscles. The therapy does not focus much on psychological causes or delve deeply into past relationships with parents. Therapy is also used to treat infants and children. The removal of early armoring allows the child to develop with a natural energy flow, emotional aliveness, and a sense of well-being. Treatment helps to prevent the development of armor in adult life.

It is important to note that the techniques used in this therapy are not the same as those in some other methods used to address physical tension such as acupressure, massage, and deep breathing. Character analysis also has its own specific techniques, which should not be confused with positive affirmations, guided self-examination, and similar therapeutic approaches.

In the Journal of Orgonomy *(Vol. 28, No. 1), Charles Konia describes a representative course of orgone therapy:*

> *In medical orgone therapy, armor is intentionally dissolved. This invariably brings about anxiety, because the very function of the armor is to prevent the (individual) from experiencing such painful feelings. The medical orgonomists encourage the individual to ex-*

perience and tolerate anxiety so that the underlying, contained emotions can be felt and then expressed. This brings about the desired, positive, therapeutic effect: anxiety is eliminated and replaced by a sense of pleasurable well-being.

The Benefits of Medical Orgone Therapy

Medical orgone therapy is a unique approach to the prevention and treatment of a wide range of mental and physical conditions. Medical orgonomists report successful treatment of the full range of emotional symptoms and relationship problems. Serious conditions such as depression, schizophrenia, panic disorder, and ADHD can often be treated without resorting to medications. Reich concluded—and present-day physicians who practice medical orgone therapy concur—that the elimination of armoring, in and of itself, restores natural, healthy functioning. Patients regain their natural capacities to enjoy satisfaction in love, work, and the pursuit of knowledge.

The Forer Effect

One day in 1948, Psychologist B.R. Forer gave a personality test to each of his students. Unbeknown to them, he completely ignored their answers and gave each student the evaluation given below (no student being aware that the same evaluation was given to each student).

> You have a need for other people to like and admire you, and yet you tend to be critical of yourself. While you have some personality weaknesses you are generally able to compensate for them. You have considerable unused capacity that you have not turned to your advantage. Disciplined and self-controlled on the outside, you tend to be worrisome and insecure on the inside. At times you have serious doubts as to whether you have made the right decision or done the right thing. You prefer a certain amount of change and variety and become dissatisfied when hemmed in by restrictions and limitations. You also pride yourself as an independent thinker; and do not accept others' statements without satisfactory proof. But you have found it unwise to be too frank in revealing yourself to others. At times you are extroverted, affable, and sociable, while at other times you are introverted, wary, and reserved. Some of your aspirations tend to be rather unrealistic.

He then asked each student to assess the evaluation on a scale from 0 to 5 ("5" indicative of an excellent" assessment and "4" meaning the assessment was good, "3" meaning fairly applicable, and so forth. The class average evaluation was quite favorable—4.26. Students were stunned, however, when they found out what really occurred. This is known as the "Forer effect".

What the Forer effect shows is that people have a tendency to see vague and ambiguous particular claims that could easily be applied to anyone (of the sort customarily found in horoscopes) as true of themselves, so long as these claims are consistent with their self-perception. In effect, what Forer showed was that there is not so much variation in how each person perceives himself and that most people see themselves in a fairly favorable light, so to speak.

The Forer effect goes some way toward explaining why people are so fond of sham sciences like astrology, palm reading, and fortune telling. Carefully conducted scientific studies show plainly that these pseudo-sciences are poor measures of personality assessment. That notwithstanding, people still flock to these sham scientists for insights into their personal problems and suggestions on what the future holds for them.

Why do they keep coming back? Forer simply believed that humans were gullible. Hence, the Forer Effect is sometimes referred to as the Barnum Effect ("There's a sucker born every minute"). Psychologists today offer less pessimistic explanations on behalf of people who frequent psychics—including wishful thinking, sanguineness, vanity, and a general human tendency to make sense out of personal experience. In doing so, clients seem to proportion their belief in the truth of a psychic's claim to their desire that such a claim be true, instead of to evidence on behalf of the claim—especially if the claim is self-flattering.

Assessment of Carl Jung and Number Dreams

Model/Hypothesis

Prediction

Background Information/Research Methods

Data

Analysis

Conclusion

Assessment of Psychics on Reading People's Minds

Model/Hypothesis

Prediction

Background Information/Research Methods

Data

Analysis

Conclusion

Assessment of Medical Orgone Therapy

Model/Hypothesis

Prediction

Background Information/Research Methods

Data

Analysis

Conclusion

Key Terms

free association
orgone energy
cold reading
Forer effect

Text Questions

* In his analysis of the extramarital-affair dream of a mid-aged man, has Jung at least given a preliminary analysis that luckier investigators may build upon at a later time, as he claims he has given, or is his interpretation hopelessly confused?
* In the example of cold reading above, does the exchange between the psychic P and her client C show that P has psychic abilities or are there more economical ways to explain the "predictive successes" of P's claims?
* What are some of the claims Reich makes on behalf of orgone therapy? To what extent are these claims scientific?

Text-Box Questions

* What is free association for Freud? Why does Grunbaum think that free association is not a scientifically valid method?
* Why are psychics' clients generally very tolerant of failures with their readings? Explain what "cold reading" is in your answer.
* What is the Forer Effect? Why is it sometimes called the "Barnum Effect"?

---

1 C.J. Jung, *Dreams* (Princeton University Press, 1990).
2 Peter A. Crist and Richard Schwartzman, *The American College of Orgonomy*, http://www.orgonomy.org/article_001.html.

Further Reading

Cohen, I. Bernard, *The Birth of a New Physics* (New York: W. W. Norton & Company, 1985).

Dijkstrehuis, E. J. *The Mechanization of the World Picture* (Princeton University Press, 1996)

Einstein, Albert, *Relativity: The Special and the General Theory*, trans. Robert W. Lawson (New York: Bonanza Books, 1961).

Gauch, Hugh G. *Scientific Method in Practice* (Cambridge University Press, 2002).

Giere, Ronald, *Understanding Scientific Reasoning*, ed. four (New York: Hartcourt Brace College Publishers, 1997).

Gower, Barry, *Scientific Method: A Historical and Philosophical Introduction* (Routledge, 1997).

Kuhn, Thomas, *The Structure of Scientific Revolutions* (University of Chicago Press, 1996).

Lloyd, G. E. R. *Early Greek Science: Thales to Aristotle* (New York: W. W. Norton & Company, 1970).

Lloyd, G. E. R. *Greek Science after Aristotle* (New York: W. W. Norton & Company, 1973).

Losee, John. *A Historical Introduction to the Philosophy of Science* (Oxford: Oxford University Press, 1993).

Machamer, Peter and Michael Silberstein (eds.), *The Blackwell Guide to the Philosophy of Science* (Blackwell Publishers, 2002).

O'Hear, Anthony, *Introduction to the Philosophy of Science* (Oxford: Clarendon Press, 1989).

Salmon, Merrilee and John Earman (eds.), *Introduction to the Philosophy of Science* (Upper Saddle River, NY: Prentice-Hall, 1992).

Schick, Theodore, *Readings in the Philosophy of Science: From Positivism to Postmodernism* (Boston: McGraw-Hill, 1999).

Schick, Theodore and Lewis Vaughn, *How to Think about Weird Things*, vol. four (Boston: McGraw-Hill, 2004).

Sukys, Paul, *Lifting the Scientific Veil* (New York: Rowman & Littlefield Publishers, Inc., 1999).

Wilson, E. Bright, *An Introduction to Scientific Research* (Dover Publications, 1991).

Zucker, Arthur, *Introduction to the Philosophy of Science* (Upper Saddle River, NY: Prentice-Hall, 1995)